OMNIPOTENT

CHALLENGE

A layperson's science of
What, How, Who and Why
YOU ARE

FRANK HAWTHORNE

This is a treatment of the greatest questions we can ask written by a layperson for lay-reading. Accordingly, this book is neither qualified nor intended as an instrument for academic study.

Contents

Preface

It is somewhat ironic that our growing scientific under-standing is having an adverse effect on our sense of purpose in human being and, consequently, our moral and social direction. Dissipation of human dignity and moral values is now evident even in our higher institutions. The object of "Omnipotent Challenge" is to re-define and re-establish this sense of purpose. It does this by offering a new and unique interpretation of the Anthropic Principle; one which includes a rationale reconciling science with the God idea without its having to relinquish its highly treasured naturalistic stance.

The work is written for lay reading. It is completely devoid of mathematics and is aimed primarily at the popular science readership as well as the wider social and scientific academia. It takes the reader, in narrative form and with illustration, through a basic chemistry of life, the Central Dogma of biology and on through the workings of star machinery to a basic understanding of thermonuclear synthesis and how this creates the ingredients of life; all the while, highlighting the wonder, the complexity and the beauty of the science together with its reflection in the Anthropic Principle.

The final chapter explains the rationale underlying the Omnipotent Challenge and describes the century in which we now live as having become the most crucial in human history in terms of our capacity either to meet its aspiration or become a self-extinguishing species. This chapter also prognosticates comparative scenarios on how this might come about depending upon the moral stance from which certain imminent and disastrous problems, as yet unmentionable in the political arena, might be addressed. And addressed they must be, the day before tomorrow, if possible. Hopefully, this small work may provide an opening platform.

In memory of my late brothers and our vibrant debates.

Tom, quite religious,

Eddie, strictly religious,

Joe, agnostic,

Larry, agnostic,

Epigraph:

If God there be then why make me; another part of Him
Suffer the man and suffer the beast; from Heaven such
Hellish sin
But should we make we to become like He from infinity of
chance
Then He and we might revel in such glorious circumstance

Chapter 1 - Some Introductory Concepts

However far we travel on our planet, whatever strange and isolated tribes of humans we encounter and going back in time almost as far as the first modern humans, some forty thousand years ago, signs of religious ritual, pageantry and monuments are eminent features of human presence. Can it be that religion is an innate appendage to human intelligence? Is it a behaviourism instilled in our genes by evolution? Is it merely a continuance of folklore born in the embers and flickering glow of ancient camp fires? If it is, it has immensely out-survived any other such in human history. If it is an evolutionary derived behaviourism what have we gained from it? How has it affected our survival capability? As well as religion being a powerful, external motivator of self- imposed disciplinary influence, one can see that it could be an instrument for social cohesion. It could be a basis upon which to declare standards, rules and laws. It could provide a leader with the manipulative power to control, set goals and maximise both individual and communal effort. Also, it could, as we see it in many situations today, provide a psychological outlet for stress, suffering, fear, failure and despair. One might well concede that, even with its many sins against humanity, there have been survival advantages in our religious tendency and that it could be an evolutionary derived behaviourism. It is more likely however that religion emerged from a coupling between deep seated traits in the human mind like fear, distrust, submissive appeasement and the newly developing traits of imagination and superstition.

Based on their scull capacity, the Neanderthals and other hominids existing coincident with early modern humans had intelligence not greatly different from that of humans today. But modern humans became modern humans because of a great jump in progressive ability beginning about two hundred thousand years ago and culminating about forty thousand years ago. This irretrievably separated them from the Neanderthals and other early hominids who, over a relatively short period of time after the emergence of modern humans, became extinct. This jump was the result of a new faculty emerging in modern humans that changed what they had in their heads from a brain to a mind. It was the faculty of imagination. The dawning of imagination within the workings of the human brain was like a first rising of the sun on a dark and bleak landscape. It was the instigation of a whole new dimension of being.

Consider a working dog lost in a severe arctic storm. If it's lucky enough to come across a shed, it will, by association, recognise it as a place of shelter. It will scrape and shove at the door. If it manages to get the door open, it will go in. But it will lie and freeze to death or confront hungry wolves before it occurs to it to try shutting the door. That's because it has no imagination. It doesn't think.

Homo erectus, who arrived on the scene some one and a half to two million years ago, is generally accepted to be our first true ancestor. He walked upright, used fire, made weapons, hunted large animals and made crude tools including hand axes.

Paleo-anthropologists can show that, over a period of one hundred and fifty to two hundred thousand years previous to the emergence of modern humans, their skulls had undergone a gradual change in shape. Human tissue of that age which might be of any paleontological use is non-existent but the interrelatedness of biological processes tells us that this gradual change in skull shape was occurring in conjunction with changes in the tissue it

2

contained. In other words, the whole was a response to genetic change over time. A gradual growth in the number and relatedness of neurons together with increased and vastly more complex connective pathways was taking place. This is consistent with the gradual improvement and refinement of tools and implements together with their decoration over that time. Brain power was improving in the direction of perceptual consciousness. Its effect was so profound as to result in its progression being categorised into Palaeolithic stages.

It is as though, by two hundred thousand years ago, evolution had reached the limit of what was achievable in the genetic transmission and expression of abstract information by means of chemically induced instinctive behaviourism and found a new and highly novel way of progressing beyond that limit. It is significant that signs of intelligence are apparent in a number of hominid species from about that time eventually culminating in the emergence of modern humans.

Evolution is driven by small, random mutations (changes) in the gene structure of organisms which are transmitted by inheritance. It is difficult for some to accept that organisms evolve progressively by means of changes which are random. However, whilst the changes in the gene are indeed random, the effects or imposition of these changes on the organism as a whole are anything but random. Some ninety-eight percent of the information in our genes rests as a kind of dormant information. It was labelled "Junk information" in 1972 by the Japanese microbiologist Susumo Ohno. This however may beg a question. Considering the amazing, apparent ingenuity of life, would it be characteristic of it to carry such a relatively enormous load of useless baggage? Fortunately, since the turn of the century, science has changed its mind on this, even though discerning the purpose of this information is proving to be rather difficult. Some recent findings would indicate that much of it codes for a kind of inter-gene communication. This indeed is most likely the case and my personal unqualified opinion is, that it constitutes a library of partial compatibilities with already

3

ongoing processes and awaits further refinement before full incorporation as an enhancement of those processes. This however, could only be witnessed over evolutionary time spans. But the gradual change in brain soft tissue configuration together with changes in bone structure for its accommodation that we have just considered would be typical of such interpolation of gradually accrued fragmentary information. Also, this answers a crucial and demanding question proffered by anti-evolutionists regarding gradation in species development. It could account for the clearly stepped changes in some morphologies apparent in the fossil record with absence of any intermediary stages.

A biological entity is a closely interrelated series of activities which combine in highly complex ways to sustain the overall form and function of the organism. A peculiar aspect of life is, that an organism has an apparent cognisance of its integrity. Changes imparted by random gene mutation are therefore gated or sorted in terms of compatibility with already ongoing processes and degree of receptivity to the functional whole. This is evidenced by things like what we have just considered i.e. changes in scull bone structure to accommodate change in soft tissue brain configuration. This is not random evolution. It is a co-relation of widely separate classes of information even though the information itself was randomly accrued. It might be likened to two engineers working in different drawing offices of a car manufacturer. One is designing a car body and the other is designing an appropriate suspension system whilst the factory as a whole co-relates them both. Also, the factory would be highly receptive to any innovation which would improve the car body or the suspension or the relationship between them. In the same way, an organism, once a graduating process has been established, becomes highly receptive to any other randomly accrued information which is conducive in the same terms. So, whilst evolution is fundamentally dependent upon randomly accrued information, the machinery of evolution i.e. life is not random. This is what is described as the "Teleonomic" quality in life. In other words, it appears as though life is following a pre-

4

designed course as opposed to the car and its suspension which have a teleological quality meaning they were the subjects of pre-design. This system is by no means fool-proof but it prevents random evolution which might produce things like wildebeest with their eyes on their backsides or dogs with their tails at the front. Having said that, a faulty mutation in a homeobox gene could cause such disfiguration but natural selection would prevent its persistence.

The soul rending wonder of evolution is, that by this means, it performs a non- cognisant, perpetual, multidirectional experimentation in structural form and function threaded to optimum practicality through consequent, self- induced natural selection. Natural Selection is an automatic consequence of the laws of physics and the way in which life works.

Like the change in a prepared recipe when it is put into an oven, this gradual change in brain configuration seems to have had a kind of coming together and culminated in a rather explosive and metamorphic change in brain power occurring about forty thousand years ago. The difference between modern humans and the Neanderthals and other coincident hominids, (Who seem to have just missed the boat), was so profound as to represent almost a difference in kind. Modern humans were now equipped to progress more per one hundred years than they had per thousand years previously. Also, their progression was now in a state of continuing acceleration. Through the faculty of imagination, they could now think. They had acquired introspective consciousness. They could have ideas and speculation. They could relate their inner being to the external world.

This book is not about life, evolution or science. (I should make it clear from the start that I am not a scientist). It is about the effects of ever-increasing scientific understanding, interpretation and speculation on our sense of human purpose and consequent social direction. Charles Darwin disqualified the biblical stories but not the God idea. We shall see in this little work that neither does modern science.

Whether the standpoint from which you consider this debate is religious, secular or atheist, it is necessary to have some understanding and appreciation of the science from which the debate emerges. Unfortunately, this would require the conveyance of several libraries of highly technical information which is far beyond the scope of this work. I shall attempt however to provide a brief overview of some of the elementary basics in the hope of conveying at least some appreciation of the wonder, complexity and beauty of the science as well as its failure to explain the apparent miracle of life.

You must, from time to time, have asked yourself the question, "What is life"? You are by no means alone in that. Science, as yet, can give no clear definition of what life is. There are many definitions of course, but all of these, without exception, describe what life does rather than what it is. One can describe what a steam engine is made of and what it does and how it does it, but that's a far cry from how it came to be and why; what motivated the design, manufacture and assembly of its components and, most interestingly, its sphere of utility.

Amazingly, of the ninety-two naturally occurring elements, almost the entire mass of everything that lives is made up of only twelve of these. In order of atomic weight from lightest to heaviest, these are hydrogen, carbon, nitrogen, oxygen, sodium, magnesium, phosphorous, sulphur, chlorine, potassium, calcium and iron. Of these, carbon, because of its particular electron configuration, is the key element. This, in conjunction with a peculiar anomaly in the way the two hydrogens set themselves about the oxygen atom in a molecule of water, forms the main basis on which life systems work. From these few elements, life constructs a mind-boggling array of substances which interact to produce an equally mind-boggling realm of form and function.

There is much talk nowadays about the building blocks of life. There is also much confusion in the lay world regarding what they are. Don't be concerned about this. It depends on the perspective from which you might be looking at life. Many botanists, for example, think of cells as being the basic building blocks. This is because they tend to see life as being built from cells. But there are many single cells that are complete living entities and must therefore be built from something smaller. Many astronomers are excited by the discovery that some amino acids can form in space. Amino acids are the building blocks of proteins which are the substance of life so they now say the basic building blocks of life can form in space. A microbiologist thinks of nucleic acids as being the basic building blocks because they are the initiators of life but they are comprised mainly of protein and so it goes on. I think it would be more appropriate to refer to these substances simply as organic compounds. They are chemical compounds which some strange facet of life puts together in the most astounding ways to construct living things which reproduce to make more living things almost identical to themselves. Bricks cement and timber are the building blocks of houses but no amount of them in any conceivable circumstance will ever put themselves together to form a house and a house will never produce another house. Nor has science, as yet, managed to put the so-called building blocks of life together in such a way as to come alive. In fact, the best scientific observation thus far seems to show life as being an extraneous force or principle which applies itself, like the builder of houses, to suitably qualified inanimate matter (The organic compounds) and sets it on an enigmatic course of evolving complexity with the sensitivity we call alive. Science rejects the idea of any extraneous force or principle and hopes one day to extrapolate innate properties in matter which account for the phenomenon we call life. I believe and hope science will be successful in this. It would be every bit as spectacular as finding ET. But for the present, the formation of the template of life remains a mystery.

The most fundamental aspect of what we call life is the capacity of a complex chemical compound to react to its environment in such a way as to replicate itself more or less exactly and to assimilate this process as a forming function in sustaining the process. Recently, a number of scientists have been successful in constructing the first self- replicating molecules outside the sphere of life. But these molecules do not adopt any form or function. They simply represent a more concentrated form of goo than the carefully prepared goo in which they replicate.

Prokaryotes and archaea are the simplest living organisms. They are the most proliferate form of life on earth. They exist as single cells and infiltrate and cover your entire body as you read. By technical perspective, science's newly engineered self-replicating molecules are to the prokaryote as a safety pin is to a jet fighter or passenger aircraft; such is the complexity of the simplest living cell. Nevertheless, the success by scientists in engineering a self- replicating molecule is no small feat. Rather, it is an outstanding and landmark achievement in our pursuance to understand the strange phenomenon of life.

Life is so encompassing of our everyday existence that we cannot help but take it for granted. Yet, even a cursory consideration of it is an intellectual adventure into a sphere of scintillating, heart moving, exhilarating wonder; a reward granted us by the diligent, disciplined, hard and often unrewarded work of many scientists in many different spheres of endeavour.

One of these, which at first sight might seem quite unrelated, is The Law of Thermodynamics; a highfaluting technical term for a concept that has deep and far reaching implications for everything that happens in our universe. It is to do with the conservation of energy. Fortunately, for our purpose, we only need to scratch the surface in order to appreciate one of the spectacular curiosities of life. The First Law of Thermodynamics used to state that "Energy can neither be created nor destroyed". When Albert Einstein showed that matter can be changed into energy and vice versa, the statement was changed to "The sum of mass and energy in the universe is constant".

The First Law is directly related to eventuality or what is known as The Principle of Cause and Effect. Every event has a cause and an effect and involves a transfer of energy. Life is a constant course of eventuality wherein environmental substances are absorbed and chemically rearranged into a vast range of substances that occur nowhere outside the sphere of life. Every such chemical rearrangement involves a transfer of energy.

Just as your car needs petrol to run and your boiler needs gas to fire, so all life needs a reliable source of energy to survive. The energy source upon which life survives is almost a closed circuit. Life energy comes from food and almost all of the food that almost all living things consume is manufactured within other living things. It is interesting to note that the petrol driving your car also comes from living things. But the circuit is not quite closed. It has a starting point which is green plants. Radiant energy from the sun is captured by the chlorophyll in green plants and, by the living chemical process of photosynthesis, is converted into plant substance. This is then consumed by herbivores which in turn are consumed by carnivores including ourselves. We are amongst the relatively few species who consume both animals and plants.

The Second Law of Thermodynamics is a bit trickier to get a hold of. It is to do with the persistence or usability of energy and its tendency towards irreversible dispersal into ever widening and more unusable states. It states that "The entropy of an isolated system continually increases". So, what exactly does this mean? What is entropy and what has it to do with life? Even an elementary treatment of entropy would require some book volumes of information and would probably lead to more confusion than learning. It is one of the hurdles for many young students of chemistry and thermodynamics. Rather than try to explain it therefore, I shall describe one or two examples in an attempt to admit the reader's mind to one of the realms of thought in which life appears to be miraculous.

Neither energy nor entropy is a substantial entity. They are abstract notions of certain attributes of some aspects of our universe. Nevertheless, they are real and affect every aspect of your existence.

Energy exists in a number of different forms and tends to change from one form to another during the process of normal usage. Some examples of energy forms are gravitational energy, potential energy, elastic energy such as in a spring, kinetic energy, light energy, stored energy, heat energy, chemical energy, sound energy, nuclear energy and so on. Coal is an example of stored energy where the energy is locked in its chemical bonding. In a coal fired power station, the coal is ignited, (Brought to a self-sustaining combustion temperature by the application of a small activation energy.) in a furnace which is supplied with a forced air draught. The fuel is oxidised from the air in the forced air draught. What is called an exothermic chemical reaction takes place i.e. a reaction which releases energy. The hydrogen-carbon molecular bonding in the fuel is re-arranged with the oxygen in the air to form mainly CO_2, (Carbon dioxide) and H_2O, (Water). Energy is released in the reaction in the forms of light and heat which is the actual fire in the furnace. (When you see something burning, you are seeing a chemical reaction taking place. Atoms are being re-arranged, substance is being changed and energy is being released.) The heat is transferred by conduction into the water of the boiler. The temperature of the water in the boiler can only reach one hundred degrees centigrade no matter how much heat is applied. The temperature of the furnace is much higher than this so it now causes the water to change from its liquid into its gaseous state and fill the chamber of the boiler with steam which starts to pressurise the boiler. This also pressurises the water causing its boiling point to rise and its latent heat of evaporation to reduce so that all of the water is turned into steam which now becomes superheated. In a three hundred megawatts power station, the steam temperature would typically be four hundred - and fifty degrees C. The superheated steam has acquired a large amount of latent energy by reason of its effort to expand; this effort being contained by the steel structure of the boiler. The superheated steam is passed through a pipe system to a turbine where its rapid expansion is converted into rotary mechanical energy. This mechanical energy is transmitted to a generator in which copper windings are moved through a magnetic field. This changes the mechanical energy into electrical energy. In most power stations the steam exhaust from the turbine will be used for other purposes although its capacity to do work is now considerably reduced due to its expanded state and lower temperature and pressure. Eventually, the steam temperature and pressure fall so low that it converts back to water vapour and is returned via a condenser and pre-heater back to the boiler. You can see from this that a given amount of chemical energy stored in a given amount of fuel has gone through several transformations into different forms of energy. The amount of energy available electrically from the generator is considerably less than the amount of energy in the original given amount of fuel. Some of the energy was expended in the form of light. Some escaped in the form of heat convection through the flue system of the boiler. Some of it was expended as heat radiation from the boiler external surfaces. Some of it was expended in conduction and radiation of heat during its passage from the boiler to the turbine. Some of it was used in generating frictional heat in the machinery of both turbine and generator. This general dispersal of the energy, including that of doing the work, into wider and more unusable states is the entropy of the system. Note, the energy was not destroyed or used up. It was spread out and degraded in terms of its availability to do work. Entropy is a measure of the amount by which a given amount of energy is less able to do work. Also, the progression of energy to entropy is irreversible. There is no way in which the energy driving the generator can be back-wound into the molecular bonding energy of the original fuel. There is no way the ash residue in the furnace and combustion by-products will ever spontaneously reassemble into the complex molecular, energy rich substance it once was. There is a universal tendency for energy to degrade into heat and for heat to degrade in the direction of hot to cold, never from cold to hot, and to disperse into ever widening spheres of less usability in the direction of universal equilibrium. In the same way, energy degrades from order to disorder or "More probable states" This is referred to as the eventual heat death of the universe.

Coal is a highly complex molecular structure reflective of the complex protein plant structures of which it is a degraded form. As a complex rather than random structure it is representative of a certain amount of information. In its use as a fuel, as well as its energy now being dispersed into entropy so has a large part of the information it contained as a complex hydrocarbon been randomised into useless noise.

So, what do we mean when we talk of order, information and noise? We are surrounded by everyday examples. A brick displays some degree of order as opposed to the lump of clay or piles of sand and cement from which it was made. If you were to find a brick amongst a pile of rock debris you would recognise it immediately as something that resulted from the intelligent application of energy in its design and manufacture as opposed to the random forces that produced the rocks. You would recognise its form and function. You would know it was an artefact because inanimate materials do not spontaneously organise themselves in that way. It is possible you might find a rock which is exactly the same size and shape as a brick but you know intuitively that such a thing is highly improbable and you would not attempt to build anything by that means. The brick is conveying to you a certain amount of information. The rocks also of course contain some information; reflective of the random forces from which they were formed. A geologist might spend half a lifetime finding it as opposed to your millisecond in recognising the brick. A house displays a much higher degree of order than a pile of bricks and a town a much higher degree than one house. Therefore, we associate order with complexity, information and improbability. The higher the degree of order the more information it contains and the more improbable it arrived without the influence of some outside organising force or principle.

Let's have a brief look at this question of order, complexity and probability. Without getting involved in mathematics, there is a simple, classical example which explains it adequately. It is the game of coin tossing.

A penny, or one, pence is stamped on either side with what we call a head or a tail. If you flick or toss a penny spinning into the air, it will land with either a head or a tail facing upwards. The probability that any particular toss will end up showing a head is therefore one in two or a half. If you toss two pennies, four possibilities are equally probable; two heads; two tails; head and tail; tail and head. The probability of two heads is now only one in four or a quarter. You can see from this that the probability of a particular result is a half to the power of the number of coins involved. If you were to toss a hundred pennies, the probability of them all falling heads would be a half to the power of one hundred or once in a thousand million billion trillion tosses. Since amongst this unimaginably high number only one particular order is correct and all the others are wrong, you can see that any form of order arising spontaneously by chance is a highly improbable event. If instead of the pennies being stamped head and tail, they were stamped on one side only with a number one to a hundred and the goal was for them to fall face up and sequentially one to a hundred in a line, the probability, even though it could happen, would be so low that you would be safe in believing it wouldn't happen in the duration of our universe. Also, if you were to attempt this experiment, in course of several million tosses you might get a fall which seems orderly. For example, they might fall in largely alternate odd and even numbers. But this is not the goal you are seeking so although it seems orderly it is without meaning. It is important to bear this in mind when considering the life scenario. Ordered events, especially complex and sequentially ordered events which are meaningful are extremely improbable.

I trust I have now opened the reader's mind to the first elementary concepts of thermodynamics, chance and probability which is all we need here. I now beg your indulgence as I labour on in an attempt to give your mind some realistic grasp of very large numbers.

We are going to consider a particular grain of sand which, by reason of some means which I leave to the reader's imagination, is identifiable from every other grain of sand. We shall call it "Bobby's grain". You are trying to find Bobby's grain and you know where it is to within four centimetres square and four centimetres deep on a particular beach. That is about one hand-full or half a million grains of sand; assuming the average grain of fine beach sand is half a cubic millimetre in volume. I think you would agree it's going to take you several weeks or even months to sift out Bobby's grain. Now consider you are not quite as sure as you thought about the location and you have to hire a digger to take out a cubic metre of sand. There are about sixteen thousand handfuls or eight thousand million grains in a cubic meter of sand. It's now going to take you and a huge army of helpers to find Bobby's grain within your lifetime. Now consider you have no idea where Bobby's grain is throughout the entire length, breadth and depth of the beach. You are now dealing with such a large number of grains that you could employ all of the national armies of the world to assist and still know that it would be a miracle if it were found in your lifetime. Now suppose you didn't even know what beach Bobby's grain was on. It could be on any beach or desert in the world. Are you beginning to get some kind of mental picture of huge numbers? Now suppose it's not a grain of sand; it's an atom; Bobby's atom. There are approximately ten million atoms in one average grain of beach sand which is five thousand billion in just a handful. But an atom could be anywhere throughout the entire mass of the Earth. And suppose you weren't sure it was even on the Earth. It could be on the Earth or any other planet or star or even in a gas cloud somewhere in space. The number of atoms you are now considering is ten to the power of eighty. That's approximately how many atoms there are in the universe.

So, everything in our universe tends towards dissipation of energy into heat, further dissipation of heat into ever lessening states of usability, order descending into disorder and degradation of information into random noise; all in the direction of some state of highly probable, useless, universal equilibrium; all that is, except life.

14

The apparent miracle of life is that it goes completely against this grain. Life contrives to take relatively simple substances from its environment, to contain them within self-isolating yet selectively-permeable boundaries and to combine and rearrange them into such degrees of order, interacting complexity and information content as to be, for the most part, beyond our understanding. As you read and perceive what is written here, you exercise the culminate product of this incredible, natural achievement. You represent the most highly ordered and complex entity in the known universe. In saying this, I cast no aspersion whatsoever on the sciences of medicine, biology, biochemistry and associated industries. Over the past two centuries and especially the past sixty years they have laboriously extricated and amassed a tremendous knowledge and understanding of life systems and how they work. This is evidenced by the tremendous medical facility we now enjoy in the UK as a result of their work and dedication. However, the complexity of life systems is such that one of the things we discover in its study is the vastness of learning yet ahead of us. One recent example of this is that, whereas it had always been thought that life chemistry confined its activities to the interplay of electrical charges in the outer electron shells of the substances it uses, it has now been discovered that the photosynthesis process utilises quantum effects. A team of scientists led by Australian chemist Fred Scholes at the University of Toronto in Canada has discovered, that in the photosynthesis process, the phenomenon of quantum superposition is utilised. This means that a specific parcel of energy is made to exist in more than one place at the same time and has the capacity to travel along a number of different pathways at the same time. This results in almost one hundred percent efficiency in the energy use, something quite unachievable by any man-made machine or device. This is a brilliant discovery but it serves to open yet another gateway into a whole new realm of biological study. Just as Edwin Hubble's application of spectroscopy in discovering the expanding universe changed the scientific world's way of thinking and began the

15

amalgamation of physics and astronomy into the new science of astrophysics, I believe the relatively new science of biophysics might lead us to a clearer understanding and possibly even a definition of what life is.

Before moving on, I should make this phenomenon of quantum superposition quite clear, if indeed such is possible. Superposition is not a case of a particle having split or divided into a number of parts. It is not a case of a particle having duplicated or replicated into separate entities as in cell division. The same particle actually exists in more than one place at the same time but it is still only one particle. In the extra-atomic world i.e. the normal world of classical physics that we live in, everything that happens, happens in accordance with laws we can determine, understand and comprehend and so is in alignment with our natural intuitive thinking. In the sub-atomic quantum realm, there are aspects of existence which seem to incorporate dimensional differences in time and space. The effects are detectable but are not verbally explainable or even comprehensible. They are intellectually reconcilable only in terms of very advanced mathematical equations. Nevertheless, like energy and entropy, they are nonetheless real.

Perhaps the best way to get your mind round superposition is to consider that these subatomic particles have what is called wave-particle duality. This means that the entity exists as both a wave and a particle. The wave aspect radiates constantly and uniformly in all directions. But any point on the wave form you might choose is also the particle. So, the particle is everywhere the waveform is. The waveform can cover all possible pathways for the particle and, since every point on the waveform is also the particle, the particle is therefore present on every pathway but it is still only one particle.

So, does life actually contradict the second law? There are some aspects of life, especially replication and death, which are difficult to measure in terms of entropy. Also, life is just as irreversible as the second law; it cannot evolve backwards.

16

However, consider this reiteration of the second law of thermodynamics - The entropy of an isolated system constantly increases. All of the life we know of exists on Earth and the energy it uses comes from the Sun. If we exclude the Sun, life is no longer an isolated system since it has a separate energy input, which, incidentally, it uses in prodigious amounts in order to accomplish its transformations. If we now include the Sun, the solar system as a whole, like everything else in the universe, is ultimately subject to the second law. Life is an integral part of the solar system and is therefore subject to the second law in the same way. This should not be allowed to blind us however to how seemingly miraculous life is.

Chapter 2 - Climates of Thinking

Almost since the time when the dawning of imagination enabled the human thinking mind, man has associated himself subjectively with powers greater than himself. When hunting was good and productive and he had a family of sons to be proud of, he enjoyed the favour of the greater powers. When threatened by earthquake, volcano, fire and flood he feared and suffered their wrath. He conjured acts and ceremonies of appeasement and so gave birth to religion. He assigned the creation of his own being and the world in which he lived to the Gods and when stunned into abject bereavement he besought them to reward his chief, fruitful wife, son, daughter or dear friend with a new life in their greater world. By this means, he enjoyed some degree of self- induced consolation. This idea has persisted in one form or another down the millennium to the present day. It did not take long for those of politic mind to realise the manipulative power they might gain by self- election as an intermediary and so began the world of prophets, soothsayers and priests together with their embroilment in political intrigue which has steered humanity down the centuries.

Because the difference between inanimate matter and living things is so obviously fundamental and categorical, the idea that life is motivated by a force or principle which is not an innate property of matter itself is intuitively welcome in the human mind.

Little wonder then that it has persisted since prehistoric times. One of the greatest inhibitors to learning is our tendency to believe what we want to believe. It is only fairly recently and exacerbated by Darwin's revelations that modern science, though not comprehensively so, has departed the idea of material spiritual duality.

The idea of a life principle which is not innate in matter varies in detail across the spectrum of human cultures and has done from time immemorial. It is interesting to note however, that the essential idea is highly consistent throughout and even in isolated populations; telling of a God or Gods who created all things and including separately, the cosmos, earth, life generally and man. Even in the most ancient of mythologies, the separateness of these acts of creation emphasises the idea that life is fundamentally different from other things and is endowed with metaphysical spirit.

Even many of the great Greek thinkers, some of whom might be considered as "Einsteins" in their time, maintained this view of life. Socrates, who lived four hundred and fifty years before Christ and whose philosophy and scientific approach still have a profound influence today, believed in Devine intuition and immortality of the human soul. Others, including Aristotle whose works laid the foundations of Western natural philosophy, believed that life, as well as having parental birth, is a purely naturalistic phenomenon and grows directly from the earth where conditions are suitable. This became known as the doctrine of spontaneous generation. It held considerable sway in the scientific community for some two millenniums.

Judaism since the time of Abraham together with Christianity since the time of Christ and Islam since the time of Mohammed have combined in the personification and closer identification of God and the God man relationship. Between them, these three constitute the most solid religious block on earth. They recognise a similar line of prophets, advocate similar lifestyles in terms of self-obligation and agree in their essential truths including personal salvation and life after death for which we should prepare. Whatever we think of them, and even though their dogmas and power severely inhibited the expansion of scientific thought over many centuries, they were the writers and keepers of the written word. Except for the last two centuries or so, we have them to thank for our social aspiration and almost our entire history of art, philosophy, and science. Indeed, the teachings of Moses, Jesus Christ and Mohammed, stripped of their two thousand years of religious embellishment of myth, magic and miracle, remain the soundest framework for human social development yet devised.

Something worth some consideration in passing here is, who were these prophets, Abraham, Moses, Jesus Christ, Mohammed? There is no doubt they existed; sparsely scattered in time and there is no doubt they must have been very extraordinary people. In our more enlightened time, we know they were not specially sent by God. So, what made them so different from ordinary men? One can only conclude they were naturally gifted with a novel mixture of high intelligence and extra keen sense of empathy. This would engender an encompassing understanding of the world they lived in together with a driving desire for non- vengeful justice. They were evolutionary jumps, what we might now call prodigies, beginning to escape the prejudicial vice grip of animal instinctive behaviourism. In a world which, for the most part, lived under pyramids of tyranny it is little wonder they were seen as being sent by God. It is little wonder that they themselves intimately believed that they were chosen ones; chosen by God to teach what only they could see as the social dream. That the ensuing centuries would produce a flood of rabbis, popes and ayatollahs who would mix the essential social truths in a cauldron of religious mayhem and with persuasive teaching substituted by subjugation would not have been a consideration in their time. This lack of foresight, if nothing else, casts serious aspersion on there being any element of Divinity in the persons of the prophets.

Excluding immediate modernity, almost all of the people who have ever lived on earth since man's earliest beginnings and including many of the great thinkers we have record of have believed in God and in His will being the extraneous force compelling ordinary matter to life; all that is, up until Charles Darwin.

Darwin's theory of evolution, now an established fact, did nothing whatsoever to resolve the question of what life is or how it began. It did however, conflict with and irrefutably condemn the biblical story of creation. It also greatly enabled the retrospective science of palaeontology to which we owe our understanding of life in Earth's past. This, in conjunction with Hubble's discovery that the universe as a whole is also an evolving process rather than a stable state, thus providing for postulate histories of the universe, seems to have turned some people's minds away from the God idea. Since neither of these discoveries throws any light whatsoever on the questions of life and God, one might wonder why this should be. Also, it is interesting to note that the currently accepted view within the scientific community of universal beginning and evolution very adeptly describes an instant of creation. If the "Big Bang" did indeed occur, then clearly the potential for the logic describing our universe existed in its very first instant. Since information is an abstract quantity existing purely in terms of intellectual perception where was the perceiving intellect at that instant?

Modern science is vividly aware of this enigma; this blatant violation of the laws of causation, this seeming miracle. For the past decade it has been struggling to come up with a "Before the Big Bang" continuum and is awash with some fantastic ideas. Let's hope, for the sake of human society, it might admit what this little work offers into its field of possibility.

At this stage, it must appear as though I am very anti-science. In fact, nothing could be further from the truth. To my mind, science and thought are amongst the greater beauties of human existence. But I am deeply concerned with the dissipation of sense of human purpose accompanying secular expansion in Western Europe. It liquefies personal finesse, integrity and human dignity. It is a recipe for social disaster. If one were to add up all of the human force underlying all of the wars and empire expansions throughout human history, fought under God but for pure material gain, it would amount to no more than Bobby's atom in comparison to that represented by modern day frictions. The antipodal attitudes of atheism and religious fundamentalism represent a force for conflict infinitely greater than anything the world has ever seen. It is a scenario wherein faith substitutes for reason, on both sides, and has the potential to generate a social and intellectual climate within which nuclear or biological extermination of whole populations may become acceptable or even necessary. If I were to ask a western European over the age of fifty, "Where might one draw the line nowadays between what is humanly dignified and what is obscene?" They would probably answer by offering the question, "Is there any human dignity nowadays?" If I were to put the same question to someone under the age of forty, I doubt they would understand the question. Clearly, we need a new way of thinking and I believe we have now reached a stage where some establishment of the human disposition is no longer just a point of interest. It is vital in terms of our potential for social progress and long-term survival.

The transitional period leading to the renaissance and the renaissance itself were a revolution in defiance of dogmatic restriction on freedom of thought. It was also however, an acceptance of greater individual responsibility. It is also interesting to note that, whereas it defied institutional religious authority, it did not defy religion. What it did do was to facilitate the disentanglement of science and philosophy and, more productively, empiricism and rationalism. It was a kind of rebirth of the medieval alchemist's way of thinking and working but in search not of gold but scientific proof of more absolute truth. My intention in this small work, if I may be so audacious, is to develop an avenue of thinking which is acceptable to both scientist and theologian and represents a basis for human morality and social aspiration in an increasingly individualistic world. It is in this respect that we shall

22

further pursue the question of what life is and why modern science should adopt such a materialistic view.

Before going on, I would ask the reader to consider how and why the materialistic view adopted by science should spread so readily through society as a whole. Apart from the pounding it is given by people like Professor Richard Dawkins in support of his fanatical absorption of Darwinian evolution which, in fact, is quite irrelevant, people generally, and rightly so, stand in awe of science. There is no question of course that most of the apex of human intelligence is to be found within the scientific community. Many, as well as having great breadth in what we call intelligence quotient, are gifted with great mathematical competence and faculty for complex abstract thinking. However, if you should meet an individual scientist, you are likely to find him or her not greatly different from yourself. Also, many are subject to fantasies and, on occasion, gross error. Take, for example, Rene Descartes, the great French philosopher, scientist and mathematician who lived during the first half of the seventeenth century and whose works are still compulsory reading in many of our universities today. Amongst many other things, he founded analytic geometry, bridging geometry and algebra which led to the discovery of infinitesimal calculus. This was the basis on which Gottfried Wilhelm Leibniz and Sir Isaac Newton independently constructed integral and differential calculus; the most widely used mathematical operators used in science and engineering to this day. Descartes spent a considerable part of his life in contemplation of the relationship between what he called the spiritual mind and the physical senses in terms of their interaction in establishing worldly truths. He was no doubt a great intellect yet one of the conclusions he drew from his contemplation was that animals do not feel pain. Because of his great philosophical and scientific standing in his time, his conclusion gave free licence for vivisection to be carried out throughout Western Europe. It is difficult to say which is the more mind boggling, the absurdity of Descartes' conclusion or the thought of the pain and suffering of countless restrained, live animals during deep surgical incursion into their organs and brains in the name of scientific research. On a more recent note, if we consider the work of the great, British mathematician and theoretical physicist Paul A. M. Dirac, who died in 1984. In furtherance of tremendous scientific work carried out by people like E. Rutherford, W. Pauli, Albert Einstein, L.V.de Broglie, Niels Bohr and W. Heisenberg, E. Schrodinger, M. Born and others, he formalised the basis of quantum theory in terms of quantum electrodynamics. In course of his mathematical constructions, he found it extremely interesting that some of his equations were equally satisfied with two different answers; one positive and one negative. This led to the concept of antimatter, now an established fact, and also confirmed Albert Einstein's concept of bi-directional time. Emergent from this, there is now a considerable body of people within the scientific community, who hold ideas which I shall leave to the reader's own judgement as to their feasibility and comparison in terms of feasibility with the God idea. One of these is the idea that time travel might become possible. The idea of time travel is beset with paradox. The reader is no doubt familiar with the grandfather paradox as well as the question, "Why haven't we seen time travellers from the future?" Another idea, devised in some measure as a way round these paradoxes, is that of different "Time lines". This in turn requires the "Many worlds" or "Multiverse" concept. The time lines idea postulates that every possible state and circumstance of matter actually exists within an infinite frame of parallel running universes. What this means is that, across a spectrum of parallel universes, Bobby's grain of sand will occupy the same position as every other grain of sand that exists and has existed and ever will exist. Bobby's grain will be ingested by every sea creature that exists or has existed or will exist and so will every other grain of sand; all in different universes. Bobby's atom, across a spectrum of universes, will occupy the same position as every other atom in our universe. It will undergo every circumstance and be involved in every process that every other atom of our universe has undergone or will ever undergo. The

24

same goes for every blade of grass, every plant, every leaf, fish, animal, insect and microbe, every molecule of every substance, every atom of every element, every sub-atomic elementary particle, every machine invented, existing and to be invented; quadrillions of quadrillions of quadrillions of universes. You will partner every human being that has existed, existing and ever to exist. You will bear children with every member of your opposite sex that has ever existed, existing and ever will exist; all in different universes. You, at this instant, are co-existing in a countless number of universes; one for every possible change in history or circumstance imaginable and countless millions more. Also, these universes are not consecutive; they are all existing now at the same time. There are quadrillions and quadrillions of quadrillions of you at this instant. You are reigning as kings, heading governments, raping babies, lecturing at universities, eating your mother, playing the stock market, running drugs, murdering your neighbour and every other circumstance you can imagine and still more. One can see immediately that any acceptance of this idea wouldn't just break a few more rungs from the ladder of human, social aspiration and sense of purpose, it would remove the ladder altogether. "If I am raping, murdering and back-stabbing in millions of other universes then why not in this one if it meets my immediate needs?" Why should the scientist who fantasises about turning up at a medieval battle in command of a modern military armoured division care about the social portent of such contrivance? These parallel universes are inaccessible by any means and their existence cannot be proven by any form of experiment. In view of its social portent, why should anyone consider this idea worth the breath it takes to express it? In the same way as evangelist groups and other fundamentalists can interpret scripture to suit their particular whim, the parallel universe scenario is part of a particular interpretation of quantum theory. A considerable body within the scientific community gives credence to this possibility and, at the same time, believe that people who believe in God are somewhat lacking intellectually. Atheism is now fashionable in the scientific community and, if they were to think about it, just about as intellectually descriptive as wearing the appropriate T shirt at a rock concert or football match. I trust the foregoing might assist the reader in forming his or her own judgement and perspective regarding the intellectual climates imposed on society by science. Scientists are no more immune to bouts of fantasy than the rest of us. Fortunately, however, they think mostly about different things and many staggering scientific revelations started as fantasised notions.

In the foregoing, it is important to note that I am not attempting to ridicule science. Whilst the "Many worlds" idea might be interpreted in terms of comic-strip fantasy, we have already very briefly considered the strangeness and counter intuitiveness of some aspects of quantum physics. It is a subject which stretches to the limit and beyond the human faculty of intellectual conceptuality. The Many Worlds or Parallel Universe idea, though not as described in the foregoing, is part of a serious and on-going debate regarding philosophical interpretations of quantum phenomena. As well as superposition, there is the question of material duality. Not in metaphysical terms but in the sense that a particle exists in two different forms at the same time; one as a particle and one as a wave form. Also, the mere fact of human intellectual observation of a wave- particle state alters the state of the particle. There is also the question of particle entanglement which apparently conflicts with the theory of relativity insomuch as it can be interpreted to offer the possibility of instantaneous, (Faster than light) interspatial communication. Einstein himself, whilst being the first to quantize light in terms of the photon went to his death very unhappy with quantum theory as a whole in terms of the theory itself, its incorporation of Werner Heisenberg's uncertainty principle and its philosophical implications; as in his famous statement "God does not play dice". However, including further refinements by Richard Feynman and colleagues Julian Schwinger and Sin-Itro Tomonga in the nineteen sixties, quantum theory stands as a very successful model of the sub-atomic realm of being. Difficult as it is in terms of concept, it describes and predicts quantum phenomena to a degree where it is applicable in a wide range of modern, every-day facility. We are fortunate indeed to have had and to have such scientific minds, including Professor Richard Dawkins, elucidating for us the spine-tingling beauty and subtlety of nature. Later, I shall offer some serious philosophical considerations within which even the many worlds or multiverse idea is not only reconcilable with the idea of human purpose but may be a necessary and integral feature of the purpose itself.

The renaissance is widely thought of as a time when culture and art affected explosive social change in Europe. However, it was not a sudden event. It was a period of some three to four hundred years of wars, national, political, economic, cultural and religious turmoil. It loosened religion's political grip and subjected Europe to an almost pandemic-like spread of humanistic ideas. This was greatly facilitated by Johannes Gutenberg's printing press and first mass production of movable type pieces early in the fifteenth century. Gutenberg's printing press, type system and oil-based printing dyes had a similar impact in its time to that of the internet in our time. It unlocked the literary vaults of monasteries and princes and was sufficiently volumous and comprehensive in application to instigate publication in colloquial as well as the classical languages. It also led to "Right of Authorship" which placed considerable responsibility on scientific and philosophical writers. Within three decades of Gutenberg's death, Nicolaus Copernicus re-invented the heliocentric idea proposed by some of the ancient Greek and Arabic thinkers. This outraged the religious thinkers of the time who maintained that the Earth was the centre of the universe. Three decades later however, Galileo Galilei took the drawing-room toy telescope and converted it into a workable astronomical instrument and confirmed heliocentricity by observation. The biblical story was now under serious attack. The church indexed Galileo's writings as improper reading under penalty of excommunication. Galileo was found guilty of heresy and spent the last eleven years of his life under house arrest. At about the same time, the great German mathematician Johannes Kepler culminated a life's work in finally ascertaining the solar planetary mechanism; so, laying the groundwork for Isaac Newton's theory of gravitation. The renaissance was indeed a storm from which emerged a whole new climate of thinking.

Living in the age of science and technology as we do, it is difficult if indeed possible for us to imagine the intellectual outlook of pre-scientific times. All that was available to people of those times was what Rene Descartes spent a great deal of time considering, the mind and the unaided senses. Monasteries and the aristocrats of that time were well versed in the language of knowledge which was Latin and an earmark of fine education was some understanding of ancient Greek philosophy. The humanists therefore, in their departure from religious authority, tended to adopt the thinking of some of the ancient Greeks. Many followed the thinking of Anaximander, pupil of Thales and teacher of Aristotle and Pythagoras. Anaximander, whilst, like most Greeks, believing in deities, did, at the same time, spend much of his life trying to define a purely natural, non-metaphysical order for the world and the universe around it. He speculated that the first humans, being so vulnerable as babies, germinated and pupated within the mouths of large fishes. He was nevertheless, a great intellect in his time and even now is considered by some to be the father of science.

The humanist thinkers therefore further perpetuated the ideas of Anaximander who proposed that life grew spontaneously from ordinary matter. Aristotle also followed this idea that life came from parentage and also from soil and putrefied matter wherever the conditions of temperature, humidity, etc. were right. Even though they had no experience of this in humans or domesticated or farm animals, other evidence was quite compelling. Snails proliferated in gardens, worms and insects appeared in middens, tadpoles appeared in stagnant ponds and grubs arose in their thousands on rotting meat. One could not escape the evidence of

28

one's eyes. Even the great philosopher and theologian St. Augustine of Hippo, some four hundred years after Christ, conceded that God, as well as creating life in the beginning, allowed for new life by spontaneous generation. He also claimed however that ordinary matter was endowed with small, invisible seeds of life. Rene Descartes, himself a catholic, also gave credence to this idea but further emphasised that life was a purely natural process arising, without any metaphysical element, wherever the conditions were suitable. The idea of invisible seeds of life appealed to many and developed into the idea of panspermia. This idea was also initiated in ancient Greek philosophy and proposed that life existed as a universal principle in the form of atomic seeds permeating the universe. Spontaneous generation and panspermia persisted as parallel ideas for more than two thousand years.

The later years of the renaissance period saw the development of a more empiricist approach to science and one of the early practitioners was an Italian physician Francesco Redi. About the year 1668 he concocted what seems to us nowadays a very simple experiment. He built a cubical open frame and enclosed within it a hind quarter of lamb and covered the whole with a fine muslin cloth. He then observed the set-up over a period of several days. He noted that the meat remained free of any germinating life but the muslin cloth soon became covered in the small white eggs of many visiting flies. He left the experiment running until the meat reached a rather unpleasant state of putrefaction and noted that it still remained free of any visible life. He then removed the covering and saw that the rotting meat became infested with maggots within a few days. Amazingly, although Redi's work did cause some serious concern in some quarters, he did not receive any great acclaim for his findings. Nor indeed did he himself fully appreciate the implications and he continued to believe in spontaneous generation.

In the early sixteen fifties, a young Dutchman, Antoine van Leeuwenhoek, was working for a Scottish cloth merchant in Amsterdam and was introduced to an instrument for examining cloth quality. It was comprised of what we would call nowadays, a magnifying glass. It was mounted on a small stand and was capable of magnifying to a power of three. Leeuwenhoek was a man of keen and enquiring mind and was so fascinated with the instrument that he made a point of acquiring one for himself. In sixteen fifty-four he left Amsterdam and returned to his home city of Delft. He married the daughter of a cloth merchant and set up his own very successful business as a linen draper. He had a rather tragic home life with four of his five children dying young; an unfortunate but common circumstance in those days. However, he soon became a very eminent citizen, acquiring several positions with the city authorities and, with further education, graduated as a surveyor. However, his primary interest still lay in his fascination with the optical effects of glass and his enhanced financial disposition allowed him to set up his own private workshop. He experimented and familiarised himself with the handling and manipulation of glass. One day in the course of this, he took a pencil-sized rod of soda lime glass and held the middle of it over a hot flame. As the middle of the rod heated towards melting, he gradually pulled both ends apart presumably in an attempt to ascertain the maximum state of elasticity obtainable. He pulled both ends of the rod stretching it in the middle until it came apart leaving two long, thin, sharp needles of glass. He then discovered, by re-inserting the thin needle of glass into the flame, that it quickly melted and fell away in small droplets. By catching these in a beaker of water, he succeeded in producing a range of different sized, perfect glass spheres from which he could make powerful lenses. He continued to experiment and improve his methods until he produced the first ever practical microscope. Some of his instruments were capable of producing magnification to the power of five hundred. Leeuwenhoek was sufficiently curious and self-disciplined to use his microscope in a strictly scientific manner. He examined a wide range of substances and astounded the whole world by his findings. He became the person to know even amongst the aristocrats of Europe. He discovered, categorised and meticulously drew or had drawn almost the entire range of micro-organisms known to us today. He gave birth to the science of microbiology and became a Fellow of the English Royal Society.

The effect of Leeuwenhoek's discoveries on the philosophical community of his time was to reinforce the idea of spontaneous generation. If a flask of water was boiled for several minutes, it was found to be completely sterile. After a couple of hours however, it could be seen to be infested with microscopic organisms. It seemed the occurrence of spontaneous generation was now observable. Leeuwenhoek who was a Reformed Calvinist and believed that all life was parental continuation of God's first creation did not agree. He asserted that the life forms entered the flask from the surrounding air. It was the resurrection of an age-old controversy and was only slightly affected by a French contemporary Louis Joblot. Joblot was a Roman Catholic, mathematician and physicist who also studied optics and was keenly interested in the new science of microscopy. He conducted an experiment which was quite ingenious for his time. Joblot prepared a sterilised nutrient solution, called an infusion. It was inserted into two flasks, one of which was immediately stoppered and the other left open. Over a period of a few days, both flasks were examined. The open flask was found to be infested with microbes (Infusoria) while the closed flask remained sterile. The closed flask was then opened and it also soon became infested with micro-organisms. Joblot's work did little or nothing to affect the argument. He received no more acclaim than did Redi and the controversy over spontaneous generation continued.

By the eighteenth century, science had attained a high cultural profile in the dining rooms and drawing rooms of the aristocratic. Also, a new generation of gentlemen's clubs, societies and coffee houses was the introduction of an interim layer between the rulers and the people deriving of merchant wealth resource. Also, the world had been greatly widened by pressure of colonial and merchant endeavour. New species of plants, animals and insects were being avidly collected from the far corners of the Earth and festooned upon the intellectuals of Europe. In the midst of this there emerged one, Georges-Louis Leclerc, Comte de Buffon. He was a French mathematician, cosmologist, primarily a naturalist and adept and prolific writer. He was a great admirer of Isaac Newton and was the first to collate Newton's calculus mathematics with probability theory. He was one of the most popular writers in eighteenth century Europe and conceived, if somewhat misconstrued, of the idea of natural evolution. From his middle age to his death he was director of the then Jardin du Roi, the equivalent of Kew Gardens, which he expanded very considerably. Buffon subscribed to the idea of spontaneous generation. He believed in a universal vital principle of life with which organic molecules were imbued and, contrary to the law of thermodynamics, that the molecules of dead animals and plants re-organised themselves into new organisms. His status as a scientist, philosopher and naturalist together with his spectacular acumen for writing impacted his ideas heavily on Europe's intellectual climate. A contemporary of Buffon, John T. Needham was a Scottish catholic priest and naturalist. He was a fellow of the Royal London Society and carried out experiments similar to those of Joblot. He claimed to have witnessed spontaneous generation and supported Buffon's ideas. However, one of the most prominent scientists in Europe at that time now entered the argument. He was an Italian catholic cleric, biologist, physiologist and teacher. His name was Lazzaro Spallanzani and he was as revered in the scientific community of his time as Albert Einstein was in modern times. He was keenly sought after by most of the prominent universities of Europe and at the age of only twenty-five became professor of logic, metaphysics and Greek at the University of Reggio. Spallanzani had already made considerable inroads in the understanding of physiology. For example, he discovered that food digestion was not a simple matter of fragmentation and solution but was an actual chemical process by the enactment of the digestive juices on the food matter. His empiricist approach to science was similar to that of Leeuwenhoek, it was highly disciplined. He carried out experiments similar to those of Joblot and Needham but within a planned and prepared environment as clinically conducive as a modern laboratory; a remarkable feat for his time. In course of his experiments he laid the groundwork for laboratory sterilisation and contaminant prevention. On the basis of his experiments, he rejected the findings of Joblot and Needham both of whom argued that Spallanzani had destroyed the life seeds by his process. The controversy continued for another hundred years until the great French chemist, physicist and biologist Louis Pasteur finally put an end to it. Pasteur was first to discover the peculiarity of what is called Chirality or handedness in organic molecules. He also confirmed the germ theory of disease, immunisation by vaccination and also instigated the partial sterilisation of nutrient solutions by what we now call pasteurisation. Pasteur designed a special flask with its only opening being at the end of a goose-neck shaped extension. This allowed the vessel to be kept open yet, impervious to the entry of dust particles from the air, by reason of a liquid seal formed in the goose-neck see fig. 1. This is the exact same principle as the liquid seal that now seals your inside toilet, sinks, bath and shower from the external, smelly drainage system.

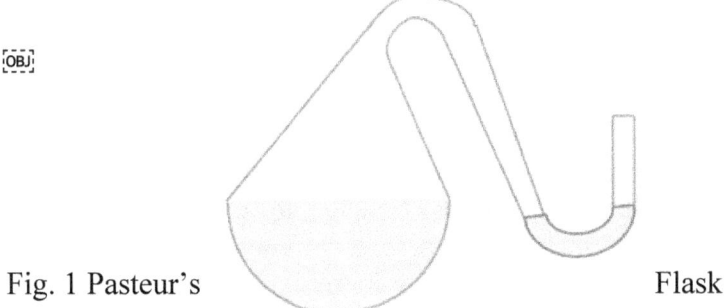

Fig. 1 Pasteur's   Flask

A nutrient solution was injected into the flask, boiled to sterilisation and left to cool. The solution then remained sterile until the flask was broken open whereupon it soon became infested with microbes. This demonstration, as well as proving germ theory, finally dispelled the theory of spontaneous generation in favour of biogenesis. The theory of spontaneous generation was never again seriously considered until, in a modified form, by modern science. The work leading up to and including Pasteur might seem like elementary school stuff to us nowadays. But at that time, airborne bacteria were as intellectually alien as particle entanglement and bent space gravitation are to us today.

The generality of opinion within the modern scientific community places all metaphysical ideas within the file marked Claptrap. But it maintains a belief in the Big Bang instant of creation, which it cannot explain. It insists that all things, including life, are explainable in naturalistic terms. Science believes that life was the subject of chance spontaneous generation but not as a general rule. Scientific evidence based on existing life indicates that it emerged from a single event of spontaneous generation or, was the sole survivor from a climate of spontaneous generation in the long distant past. It was a chance event occurring within a particular climatic window of opportunity. Scientific evidence would indicate that life began when the Earth was still very young and there has been much speculation regarding what the climatic conditions might have been like at that time. Laboratory experiments with simulated hypothetical conditions and using various lightening type energy inputs have been carried out since the early nineteen fifties. The most famous of these was the Miller-Urey experiment. This tested to see if speculated, primitive Earth conditions were conducive to chemical reactions that synthesised organic compounds. It was considered a classic experiment on the origin of life and it did produce amino acids which life uses to make proteins as well as lipids, sugars and some of the building blocks for nucleic acids. Many more experiments have since been carried out, some with even better results. Ingenious, exciting and promising as these experiments have been however, the reader should understand that the substances they have produced are no nearer being alive than earth-bound metal ores are to being a locomotive. However, the experiments do strongly support the

Oparin Haldane primeval soup theory; something we shall be considering later.

Chapter 3 - Basic Chemistry of Life

With the reader's newly re-enforced, if somewhat superficial, appreciation of thermodynamics, probability and grasp of huge numbers, we shall take a look at the kind of chance science is talking about. In order to do this, we need to take a very superficial look at life chemistry. The reader needn't try to learn or remember any of this. The mere reading and reasonable understanding of it will suffice to generate the conceptual frame of mind within which one might reasonably evaluate the subject matter. Also, I trust the presentation might be of some interest.

Our senses are tuned to survival. They therefore present a somewhat superficial view rather than an understanding of our environment. You look at a brick wall or a chunk of wood and see an apparently inert solid. You see a jug of water; it's a lot less solid but just as inert. Then you have the gasses, most of which you can't see, but can feel if they are pressurised against your body. But these apparently inert substances are in fact highly active spheres of activity. The brick wall is a vast collection of atoms which in themselves are comprised of highly interacting forces and which interact further to form molecules. These are in a constant state of violent vibration and random collision which determines the temperature of the whole. We may think of our universe as a sphere of highly interactive activity rather than substance. It is the nature of this activity that determines what we see as substance. Everything that exists, exists by reason of interacting plurality.

Atoms

There are ninety-two different, naturally occurring atoms called the elements. There are also some twenty odd other man-made atoms most of which are very unstable. Of the ninety-two natural elements, life is essentially comprised of only twelve of these, as mentioned earlier. Of these twelve, our discussion involves only a few. The gross assembly of an atom is comprised of three main types of interacting parts: neutrons, protons and electrons. These in themselves are comprised of other interacting particles so that the whole comprises what we might call an interacting force field activity. Fortunately, for our purposes, we need only deal with the superficial concept of neutrons, protons and electrons. A collection of atoms is called a molecule and if all the atoms are of the same kind it is an elemental substance. If there are different kinds of atoms combined in the collection it is a chemical compound. For our purposes, we may consider an atom as being like a small planetary system. The sun in the centre, which is called the nucleus, is comprised of protons and neutrons (All except for hydrogen which has one proton and no neutron). The protons have a positive electric charge and the neutrons have no electric charge. The electrons, which orbit the nucleus like frantic little planets, have a negative electric charge. Unlike the planets around the sun however, the electrons don't orbit on a single plane but rather circumscribe a kind of sphere referred to as an orbital or electron shell. The total number of electrons is equal to the number of protons in the nucleus and this number represents the Atomic Number in the table of the elements. For example, hydrogen is number one, helium number two, lithium number three and so on. It also determines the chemical properties of the atom i.e. the ways in which it may react with other different atoms.

Fig 2 shows an atom of hydrogen with one proton and one electron, an atom of helium with two protons, two neutrons and two electrons and an atom of lithium with three protons, three neutrons and three electrons. Almost the entire mass of the atom is

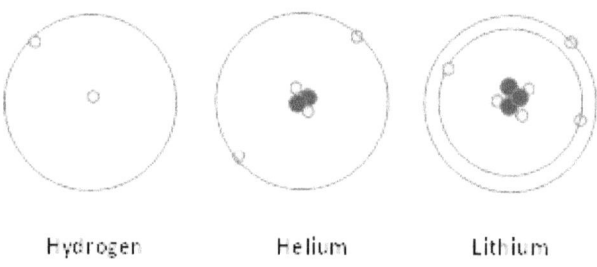

Hydrogen   Helium   Lithium

Fig. 2

concentrated in its nucleus. Like hydrogen, helium is a very lightweight gas. Lithium with only the slight difference of one electron, one proton and one neutron, is a bright, silvery, low density metal. You can see from this that relatively minor differences in atomic structure (Or force field configuration) can represent what we see as very categorical differences in substance. The electrons orbit the nucleus at discrete distances each of which represents a different energy level. The outermost electrons have the highest energy levels. Also, the orbital is not a single line. It has a depth and is called a shell. If you were to take a child's football and suspend it in the centre of a larger football the space between the inner surface of the larger ball and the outer surface of the smaller would be like an electron shell. Also, the shells themselves are divided into sub-shells. In order to move from one energy level to another an electron's energy has to be altered by a very specific amount called a quantum. Each electron orbital is capable of containing a maximum number of electrons in accordance with the simple formula 2(N squared) where N is the orbital number starting from the innermost. In Fig 2 you can see that the hydrogen atom has one orbital containing one electron. Helium has one orbital containing two electrons. In lithium, the first orbital has two electrons which fill it and a second orbital is started containing the third electron. This second orbital is capable of containing up to eight electrons, (Two squared equals four by two equals eight). At this point, including the two electrons in the first orbital, the total number of electrons would be ten and the substance would be the noble gas neon. It is important to note that 2(N squared) denotes the maximum number of electrons an orbital can contain but the total number of electrons in an atom of any element is determined by the number of protons in its nucleus. As a result, most atoms have less than the maximum number in their outermost orbital. As you shall see shortly, this is most fortunate because otherwise we would not be here. Also, whilst the number of orbitals in the heavier elements goes up to seven, (seven squared equal's 49 by 2 equals 98) no known element has an orbital containing more than 32 electrons. Because the total number of negatively charged electrons balances the number of positively charged protons in the nucleus, an atom is normally without electrical charge. It is said to be neutral. If an atom gains an extra electron it becomes negatively charged. If it loses an electron it becomes positively charged. In such situations, the atom becomes a negative or positive ion. There are many more aspects of atomic configuration and categorisation which, fortunately, do not concern us here.

Linus Pauling: courtesy Wikimedia commons

During the early nineteen thirties, the great American scientist Linus Pauling, the only person ever to win two unshared Nobel Prizes, defined the means whereby atoms form molecules. I mention this in tribute though our treatment of that subject here will be as superficial as saying "The story of the three bears is about three bears."

Only an atom's outermost electron orbital is involved in the science of chemistry. Although atoms are electrically neutral and stable in their natural state, this doesn't mean they are perfectly happy. An atom is only perfectly happy when its outermost orbital contains two, eight, eighteen or thirty-two electrons such as in the noble gasses all of which have eight electrons in their outermost shell. Whilst this statement is not completely true and is further complicated by sub-shells within the shells, it suits our purpose here and this tendency of atoms to achieve this happy state facilitates molecular bonding. This is the force underlying the whole science of chemistry. The outermost orbital of an atom is called its valence shell. Elemental atoms of different kinds combine in two main ways in order to achieve their happy state and, in the process, form chemical compounds. One is the taking and giving of electrons between atoms and this is called ionic bonding. The other is by atoms sharing electrons and this is called covalent bonding. There is another way to bond which is called hydrogen bonding and this is crucial to life.

Chemical Compounds

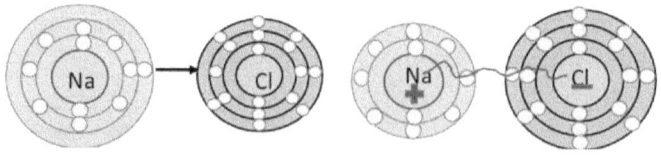

Fig. 3 Sodium and Chlorine to Sodium Chloride

A classic example used in many elementary chemistry courses to illustrate ionic bonding is that of sodium and chlorine. Sodium is a highly reactive, silver coloured metal. It has two electrons in its first orbital, eight electrons in its second orbital and one electron in a third orbital. It reacts so violently with water that flames are produced when it gets wet. Chlorine is a green coloured, highly poisonous gas. It has two electrons in its first orbital, eight in its second orbital and seven in its third orbital. When these two elements combine the single electron in the third orbital of the sodium is given off and incorporated into the third orbital of the chlorine. This results in both atoms achieving their happy state of eight electrons in their outermost shells, as in Fig. 3.

The sodium atom on the left loses its single valence electron to the chlorine atom on the right. The sodium atom has become a positive ion and the chlorine atom a negative ion.

The combination of these two highly volatile elements produces the benign chemical compound sodium chloride which is common table salt. Here again one can see that the change in disposition of only one electron in the combination of these two elements produces a categorical change in substance. It is also worth noting that the electron shell in which the outer electron of the sodium atom operated is transferred with it to the chlorine atom so that the sodium atom becomes smaller and the chlorine atom larger. The shell therefore is not just a spherical space within which the electrons operate. The shell is part of the overall force field configuration or waveform of the electron itself. The chlorine atom is so covetous of its newly gained electron that it holds onto it fiercely. If sodium chloride, or table salt, is immersed in water the respective atoms will separate easily into sodium and chlorine ions but as ions, they have lost the fierce volatility of their original atomic states. They are now parts of a simple table condiment. Ionic bonding is common to all metals and they tend to form when the numbers of their valence electrons (Outer shell electrons) are very different, like the metal sodium which has one and the gas chlorine which has seven. When an atom is in its happy state of having two, eight, eighteen or thirty-two electrons in its outermost shell it is said to be saturated.

The other and more common way in which atoms combine into chemical compounds is by covalent bonding. This occurs mainly in the non-metal range of substances. Covalent bonding is achieved by the pairing up of one electron in an atom with one electron in another atom and this pair being shared by both atoms in attempting to saturate their outermost shells. Some atoms can contribute more than one electron so that multiple pairs can be formed and shared with other atoms in forming more complex molecules. Carbon, the key element to life, has two orbitals. The first orbital has two electrons which saturate it and its outer orbital has four electrons which place it mid-way between zero and saturation and also mid-way between metal and non-metal. It will combine with elements of either group. Pure carbon occurs as a compact tetrahedral crystal in diamond, as thin, flat, layered crystals in graphite and as a black, slippery powder.

We are going to take a brief look at some carbon chemistry. As mentioned earlier, the reader needn't try to learn or remember any of this. But carbon is the key to life and what is important for the reader to note is the way in which carbon molecular bonding facilitates gradual change in substance and properties. This makes it highly conducive to the sensitivities and lability of life systems. Also, as you will see, it lends itself readily to the building of volumous and complex molecular structures. Fig four shows an illustration of the carbon atom but this is not what an atom actually looks like. As mentioned earlier, an atom is a constant, extremely rapid, complex, highly interacting force field configuration. It cannot be illustrated and can be understood only in terms of mathematical equations. However, Fig. 4 is a typical illustration which allows us to visualise how valence works in combining atoms into molecules. Any other illustrations or forms of notation should be seen in the same light.

Fig. 4

Each of the four valence electrons (Outer electrons) in Fig. 4 can be paired up with a valence electron of another atom or atoms and shared as a pair in contributing to the saturation of the outer shells

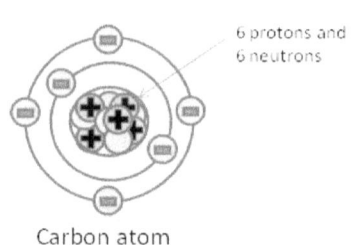

6 protons and 6 neutrons

Carbon atom

42

of the participating atoms.

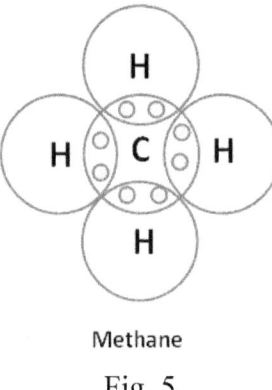

Methane

Fig. 5

Fig. 5 shows a carbon atom in the centre with each of its four valence electrons paired up with the valence electron of a hydrogen atom. By pairing and sharing their valence electrons in this way, all five atoms have achieved their happy state of saturated outer electron shells; two for each of the hydrogens and eight for the carbon. This particular combination produces a molecule of the gas methane, CH4 which is the main ingredient in natural gas. At this stage you might ask "Have the hydrogens now changed into helium"? No, they haven't because their nuclei remain unchanged.

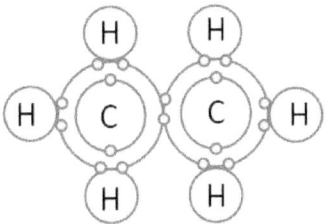

Fig. 6 Ethane

In the colourless, odourless, flammable gas ethane, C2H6, in Fig 6, you can see that carbon atoms can pair up and share their outer electrons with other carbon atoms in forming covalent bonds.

This allows carbon chain and other more complex molecules to form.

Now that you have some kind of mental picture of how atoms combine, we shall go from pictures to a more formal notation. This uses the symbols for the elements from the official table of the elements. It includes the symbols connected by a straight line or lines indicating the covalent bonds. On this basis, the molecule of methane CH4 shown in Fig. 5 would be shown thus:

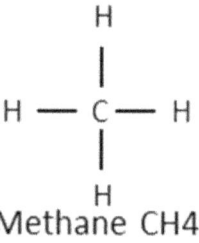

Methane CH4

[OBJ] Hydrocarbons.

The following shows a basic hydrocarbon series where, starting with ethane, the carbons themselves are covalently bonded as well as being covalently bonded to hydrogens.

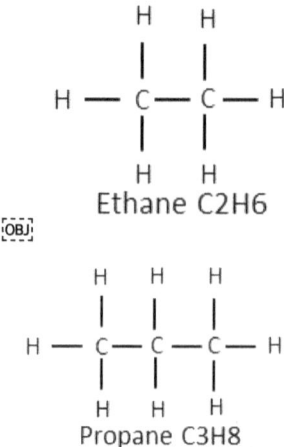

Ethane C2H6

[OBJ]

Propane C3H8

44

```
    H    H    H    H
    |    |    |    |
H — C — C — C — C — H
    |    |    |    |
    H    H    H    H
```
Butane C4H10

```
    H    H    H    H    H
    |    |    |    |    |
H — C — C — C — C — C — H
    |    |    |    |    |
    H    H    H    H    H
```
Pentane C5H12

In this series you can see a gradual change in both substance and chemical properties; a characteristic so necessary to living systems. Each of these compounds is equivalent to its preceding one plus one carbon and two hydrogen atoms. In every case, all of the carbon bonds are fully engaged. The properties of the compounds change progressively with increasing size of the molecule. Methane is the principal ingredient in natural gas. It liquefies at minus two hundred and-fifty-eight degrees Fahrenheit. Propane and butane are also derived from natural gas or petroleum. They are used as bottled gas and can be liquefied by moderate pressure at room temperature. The carbon atom facilitates an even more gradual and subtle change in chemical properties. If, for example, we take the molecule of butane, the structure of its formula can be re-arranged to produce isobutane.

```
      H   H   H
      |   |   |
  H --C --C --C--H
      H   |   H
      H --C--H
          H
```

Isobutane

This is still a saturated molecule with four carbon and ten hydrogen atoms; but it is not the same substance as normal butane. This re-arrangement of its atoms gives it slightly different properties. These types of molecules which have the same atomic formula but structured in different ways are called isomers. The number of possible isomeric variations increases rapidly with the number of atoms in the molecule. Butane has two isomers, pentane has three. The molecule with twenty carbons can be arranged in over a quarter of a million different ways. Isomers provide a means of even more sensitive gradations in properties than developing molecular series.

One can already see the incredible range of possible variation and complexity facilitated by the combining properties of carbon. Yet we have not even begun to see just how extensive it is.

Methane is the simplest hydrocarbon and if one or more of the four hydrogens is substituted by another atom or atoms it becomes what is called a substituted hydrocarbon. Substituted hydrocarbons clearly illustrate the immensity of variation in both substance and chemical properties, so essential to life, which is facilitated by the electron configuration of carbon. Chlorine is one of a class of elements called the halogens. All of the halogens have seven electrons in their outermost shells. They need only one more electron to reach their happy state of saturation and are therefore highly reactive.

Starting with methane, we shall, one step at a time, substitute one of the hydrogens with an atom of chlorine and see the gradual change in both substance and chemical properties that results.

$$H-\overset{\displaystyle H}{\underset{\displaystyle H}{\overset{\displaystyle |}{\underset{\displaystyle |}{C}}}}-H$$

[OBJ] [OBJ] Methane

$$H-\overset{\displaystyle H}{\underset{\displaystyle H}{\overset{|}{\underset{|}{C}}}}-Cl$$

[OBJ] Methyl chloride

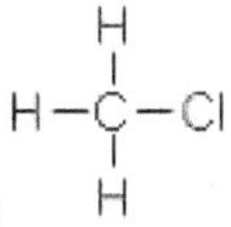

$$H-\overset{\displaystyle Cl}{\underset{\displaystyle Cl}{\overset{|}{\underset{|}{C}}}}-H$$

[OBJ] Dichloromethane

$$Cl-\overset{\displaystyle Cl}{\underset{\displaystyle Cl}{\overset{|}{\underset{|}{C}}}}-H$$

[OBJ] Trichloromethane (Chloroform)

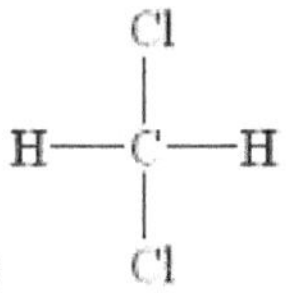

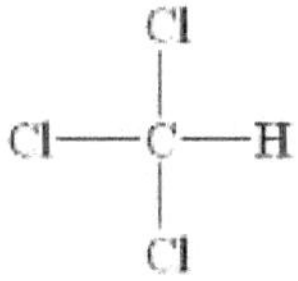

$$Cl-\overset{\displaystyle Cl}{\underset{\displaystyle Cl}{\overset{|}{\underset{|}{C}}}}-Cl$$

Tetra chloromethane (Carbon tetrachloride)

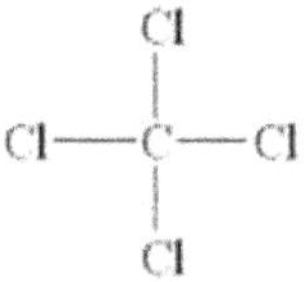

Again, in this series, we see a gradual change in properties from the highly inflammable gas methane to the non-inflammable heavy liquid carbon tetrachloride so useful as a cleaner, de-greaser and fire extinguisher. Similar series of substituted methane compounds are formed with bromine, iodine or fluorine. Similar substitutions of chlorine and related atoms can be made in the other members of the methane series, producing more varied substances.

Instead of adding a chlorine atom, another substituted methane series can be formed by adding an oxygen atom. The oxygen atom has six electrons in its outermost shell so it has a valence of minus two. In this substituted methane series one of the two valence bonds of oxygen is engaged with a hydrogen and the other bonds the oxygen to a carbon atom.

```
         H
         |
H ——— C ——— O ——— H
         |
         H    Methyl Alcohol
              CH3OH
```

```
    H     H     H
    |     |     |
H —— C —— C —— C —— OH
    |     |     |
    H     H     H    Propyl Alcohol
```

```
    H     H     H
    |     |     |
H —— C —— C —— C —— H
    |     |     |
    O     O     O
    |     |     |
    H     H     H
```
Glycerin C3H5(OH)3

So, the oxygen-hydrogen combination, called the OH group, behaves as a single atomic unit with a valence of one. The substances that are formed by the substitution of an OH group in hydrocarbon molecules are the alcohols. Also, any member of the methane series acts as a monovalent unit if one of the hydrogen atoms is removed. We can consider these combinations as being derived from a methyl or an ethyl or a propyl group. More than one OH group can be substituted in these compounds. Glycerine is an alcohol formed by propane with three OH groups.

Magic of the Benzene Ring.
There are many series of substances based on long chains of carbon atoms other than the methane series. Benzene is a hexagon of carbon atoms with a hydrogen attached at each corner. It is a closed chain and is an unsaturated compound, since each carbon has one vacant bond.

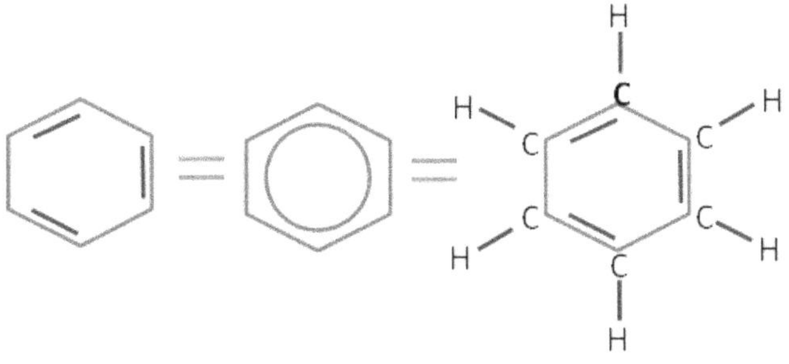

Fig. 7 Benzene

Benzene is one of a class of substances known as the aromatic hydrocarbons. They were called aromatic by the medieval alchemists because of their scents, smells, perfumes and stinks. The name has stuck but has a different and highly technical meaning in modern chemistry. In both the left and right hexagons in Fig. 7 you will note that there are double and single lines on alternate faces. This is to indicate alternating double and single covalent bonds. This should constitute an asymmetric molecule but in fact the molecule behaves as one with perfect symmetry. The middle image in Fig. 7 shows the hexagon with an inner circle. This is to indicate a rather fascinating phenomenon called resonance; a feature of the aromatic hydrocarbons. The mechanism of this kind of resonance is highly technical and any explanation would require a much deeper understanding of atomic structure than is needed in our discussion. It involves the overlapping of what are called P orbitals, their interaction with the nucleus and the activities of what are called delocalised electrons, (Although there is still some dispute regarding this particular aspect). The upshot however, is that the single and double bonds appear to change places with one another at such a high frequency that the molecule appears and behaves as one with perfect symmetry. Just like the methane series, benzene also can be modified by substituting other atoms or groups of atoms for hydrogen.

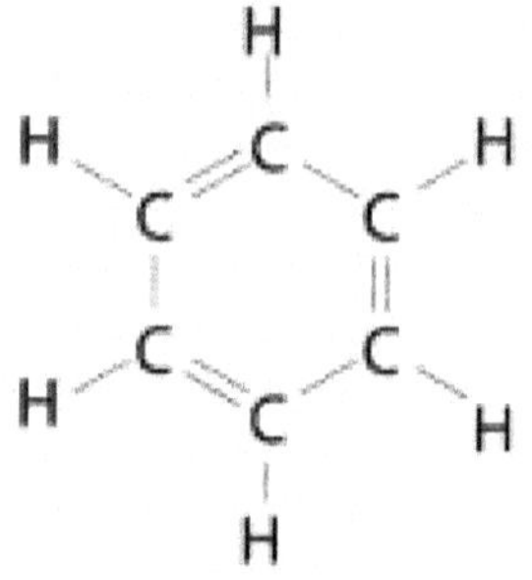

H

H   C   H
   C   C
          ‖
   C   C
H   C   H

H

Benzene

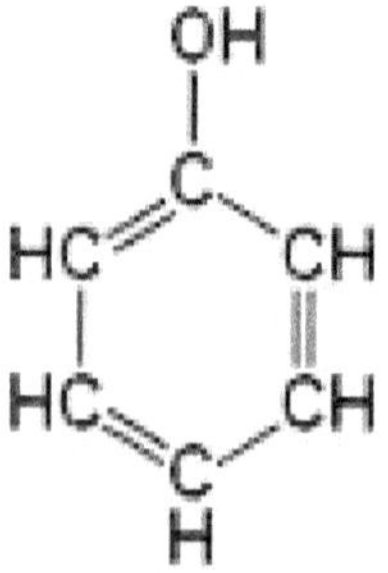

OH
|
C
HC   CH
| ‖
HC   CH
   C
   H

[OBJ]Phenol

[OBJ] Benzene is a rather nasty, colourless, highly flammable liquid. It is derived from crude oil, petroleum, coal tar and poor combustion of biomass. If one hydrogen is substituted by an OH group, the product is phenol, or carbolic acid which is a smelly, highly poisonous solid. Many such substitutions can be made including in the carbon ring itself. Replacement of one CH group by nitrogen produces Pyridine. Note the introduction of this new

50

element nitrogen. It is of great significance at the fundamental level of life systems, as is the pyridine molecule also.

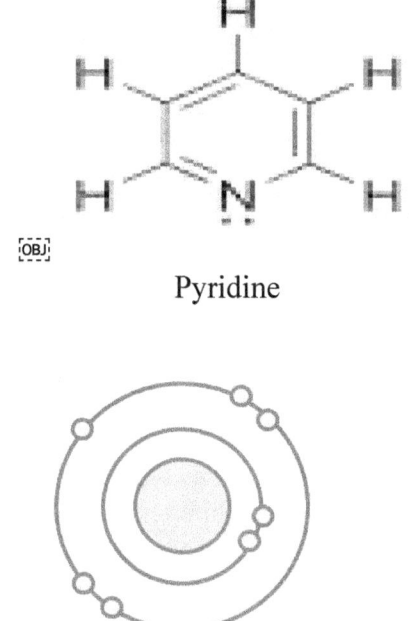

Pyridine

Nitrogen

The benzene ring can act as a unit in chains of rings and very extensive chains are possible. Every hydrogen atom in such a chain can be replaced by another atom or group and these groups may also contain substitutable members. The carbon rings can also be modified so that the possibilities of carbon chemistry are almost infinite. They are utilised with mind boggling subtlety and ingenuity by life systems.

Whilst the way in which life systems work is not part of this discussion, I think the reader might be interested in taking a cursory glance at how life manipulates the chemistry of hydrocarbons into that of carbohydrates so necessary for its accessory functions. Living organisms are comprised essentially of six classes of substances together with small amounts of a few others. These essential substances are the carbohydrates, sugars, starches, cellulose, fats, water and protein.

Sugars.

Sugar is one of life's most basic materials. Almost all life as we know it depends on the energy of sunlight. Green plants store this energy by using it to synthesise sugars from carbon dioxide and water. This is an endothermic reaction which absorbs the energy of the sunlight which then becomes available to the plant. It also becomes available to animals that feed on plants by fermentation or oxidation of the sugar.

The most common sugar deriving of green plant synthesis is d-glucose (dextrose). It is one of several isomers and is a substituted derivative of hexane, the sixth member of the methane series of hydrocarbons, with five hydrogens replaced by OH groups and two by a single oxygen atom at the end of the chain. Glucose is therefore closely related to the alcohols.

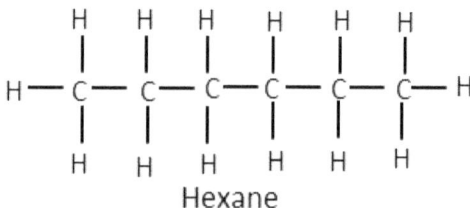

Hexane

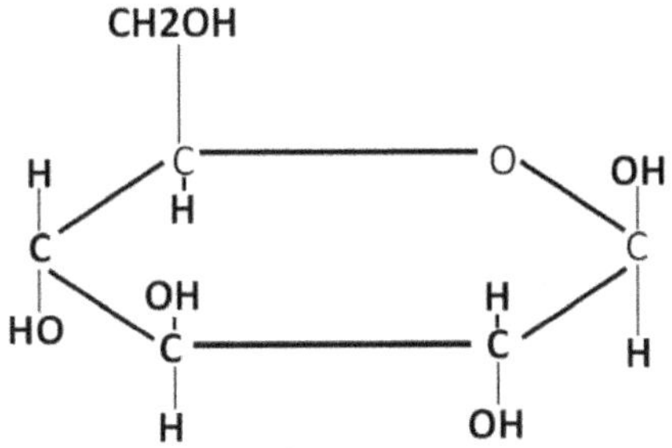

CH2OH

HO

OH

H

OH

H

O

OH

H

H

D glucose

[OBJ] Many different isomeric arrangements of the glucose structure are possible. (Remember isomers are molecules with the same atomic makeup or formula but with some of the atoms arranged slightly different). The most common isomer, d-glucose is usually drawn as a hexagonal ring rather than a straight chain. But even this single isomer of D glucose occurs in two different forms designated alpha and beta. The only difference between the two is that in one of the molecules, one CHOH group is upside-down in comparison to its counterpart in the other molecule, (Extreme right on both diagrams)

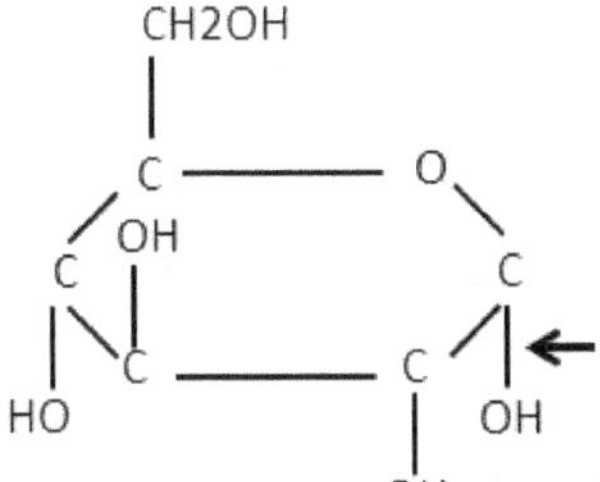

Alpha Glucose 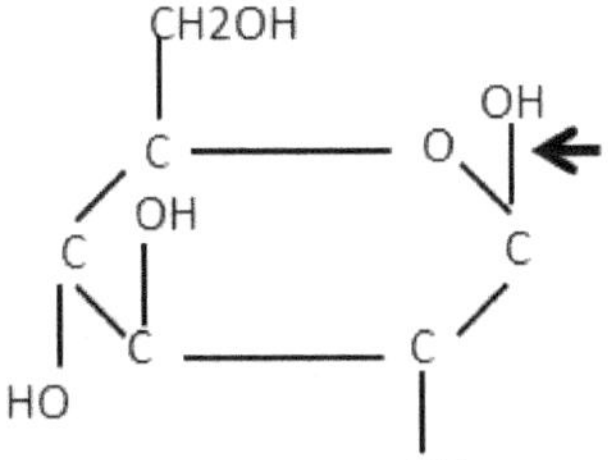   Beta Glucose

[OBJ] This seemingly slight variation in molecular configuration leads to a very important difference in properties and substance. It is the difference between starch and cellulose.

Glucose is one of the so-called simple sugars which are single-ring molecules with the formula C6H12O6. Another series of sugars is formed by the combination of two single-ring sugars into a single molecule. Two glucose molecules can be combined by removing a hydrogen from one end and an OH group from the other. The remainders of the two glucose molecules then become linked by the vacated bonds and the H and OH groups form water. This kind of linkage of organic molecules into chains, by the removal of water, is called polymerisation or condensation. The breaking apart of a polymer by the putting back of water is called hydrolysis. Both these processes are used extensively in living systems. They are chemical reactions and they don't just happen willy-nilly. Like all biochemical reactions, they are activated by highly specialised Nano-machines called   enzymes which are proteins and these in turn are specified and controlled by genes.

The double sugar formed from a-glucose is maltose. A more common type found in plants is cane sugar or sucrose which is formed by a similar polymerisation of a molecule of a-glucose with one of another simple sugar, fructose (Levulose)

Starch.

Starch is the main form of carbohydrate storage in plants and is the most important source of carbohydrate in human food. It is composed of two different molecules; (a) amylose, which is insoluble in water and (b) amylo-pectin, soluble in water. The starch molecule is a polysaccharide (A carbohydrate comprising a number of sugar molecules) assembled from the simple sugar glucose. It can be comprised of from five hundred to several hundred thousand glucose molecules joined by covalent bonds into a single structure. Because of a difference in the way the glucose molecules are linked, amylose and amylo-pectin form different kinds of structures. Amylose forms an overall spiral shape and amylo-pectin forms a tree-like, branched structure. Plant starch, on average, is twenty to thirty percent amylose and seventy to eighty percent amylo-pectin. If the long chain is made up of b-glucose units, (Remember b differs from a only in that one CHOH group in the ring is upside-down) the resulting substance is not starch but cellulose. The chemical formula is the same as for starch, C6H10O5 but in cellulose it is a much longer chain.

Fats.

As we have seen, the carbohydrates have structural formulas similar to the alcohols. Fats are comprised of the same elements, hydrogen, carbon and oxygen; but they are more complex. We shall have a very cursory look at two of these, glycerine and formic acid. Glycerine is the hydrocarbon propane with three hydrogens replaced by OH groups. Formic acid is methane with one hydrogen replaced by an OH group and two hydrogens replaced by oxygen. The inclusion of this COOH group produces acid properties in any compound in which it is present and the COOH substitution can be made in any member of the methane series of hydrocarbons. Acetic acid is produced in this way from ethane, and butyric acid from butane. Butyric acid and higher members of this substituted series are called fatty acids, because they combine with glycerine to form fats. The combination forms like the polymerisation of simple sugars, by the removal of three OH groups from glycerine and a hydrogen from each of the three acid molecules. These groups form water, and the vacated bonds of the glycerine and acid molecules link up to form a fat. Fats provide a more efficient form of energy storage than sugars or starches.

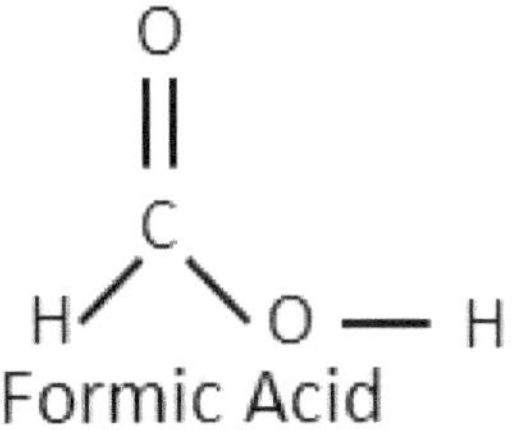

$$O$$
$$\|$$
$$C$$
$$H \diagup \diagdown O - H$$

Formic Acid

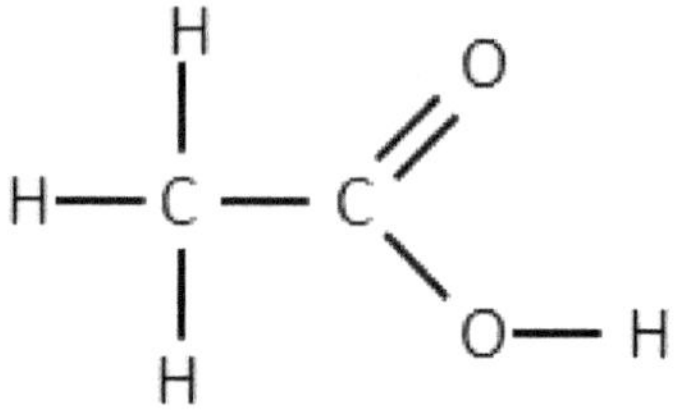

Acetic Acid

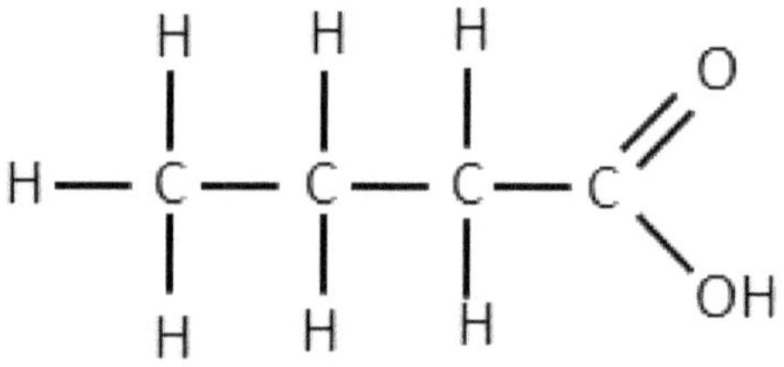

[OBJ]

Butyric Acid

[OBJ]

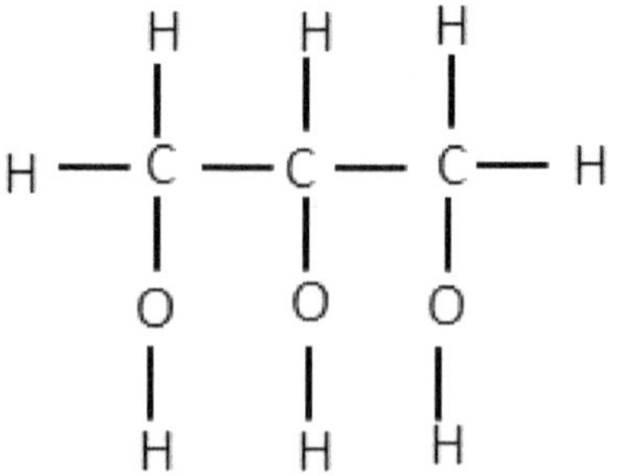

Glycerin C3H5(OH)3

Water.

Except for spores and seeds which are in a dormant, dehydrated condition, living systems generally exist within a watery medium. But water is not just a convenient medium within

56

which things can easily float around. It enters chemically into the complex, structural and functional aspects of everything that lives. Although water is a plentiful and seemingly very ordinary substance, it does, in fact, have some very extraordinary properties. Its simple chemical formula is H2O which means it is comprised of two atoms of hydrogen and one of oxygen. You will remember that hydrogen has one electron in in its single shell and oxygen has six in its outermost shell. By bonding covalently, all three atoms achieve their happy state of full outer electron shells; two for each of the hydrogens and eight for the oxygen.

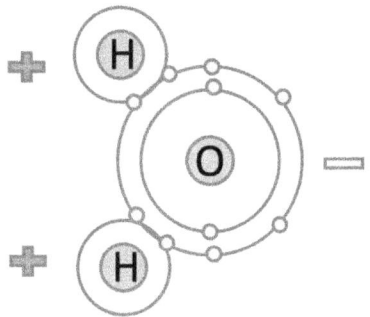

Fig 8 Water

However, whilst one would naturally think that the two hydrogens might set themselves in diametrically opposed positions one hundred and eighty degrees apart across the oxygen atom, you will note, from fig. 8, that this is not what happens. The hydrogens set themselves at approximately one hundred and five degrees apart. This results in a molecule which is lop-sided both structurally and electrically. The proton nuclei of both hydrogen atoms are left unshielded with the result that the molecule exerts a positive electrical influence on the hydrogen side which is similar to but generally weaker than that of a positive ion. This rather unique nuclear attraction is called the hydrogen bond. It is most prominent in water but also occurs in other molecules when a hydrogen atom has a prominent position in a ring or chain. In combining with the hydrogen atoms to form water, the oxygen atom now has two more electrons than it has protons in its nucleus. It therefore has an excess negative charge so that the molecule as a whole is polarised; positive on the hydrogen side and negative on the oxygen side. This is what accounts for water being so efficient as a solvent. Referring back to Fig.3, you will remember that sodium and chlorine are joined by an ionic bond to form table salt. When salt is placed in water, the positive sodium ions are attracted by the negative oxygen sides of the water molecule and the negative chloride ions are attracted by the exposed positive hydrogen nuclei. This exerts sufficient force to break the ionic bonds so that the salt goes into solution and both ions are surrounded by water molecules. Even more amazing, is the way in which water enters actively into the complex activity of living systems. The peculiar influence exerted by the exposed hydrogen nuclei is weaker than that of ionic or covalent bonds. It is therefore able to interact in the assembly of large and complex molecular structures and, at the same time, remain amenable and sensitive to small stimuli too weak to affect ionic or covalent bonds. The lability of life systems i.e. their ability to respond to stimuli is due largely to the sensitivity of hydrogen bonds. Hydrogen bonding can be a study on its own; requiring an advanced knowledge and understanding of physics, mathematics, chemistry and biology. Let us suffice to say here that its influence is exerted over a range of angles and strengths varying from weak to the equivalent of ionic and covalent bonding. Water has other exciting properties. At 0 degrees C. it has a high freezing point and a high boiling point at 100 degrees C. Since earth's surface temperatures lie between these, most of the water remains in liquid form. It has a high latent heat of evaporation which is the amount of heat needed to change it from the liquid to the gaseous state. This makes it an excellent coolant for life systems. It also has a high specific heat. This means it absorbs a considerable amount of heat for relatively small increases in temperature. In this respect it has a kind of temperature

58

inertia which protects life systems from sudden environmental changes. Because of its polar nature, water has a strong cohesion which facilitates osmosis in plant life forms. Another rather unique property of water is that its solid (Ice) state is less dense than its liquid state so that ice floats. This ensures that water freezes from the top down rather than bottom up. If it froze from the bottom up, most of the water on earth would remain frozen most of the time making it practically impossible for aqueous life forms. You will be aware that unlike electrical charges attract and like charges repel. Depending on the orientation of water molecules therefore, they also have a tendency to come apart or dissociate into hydrogen ions or what are called hydronium ions H_3O+ and hydroxyl ions $HO-$. In ordinary water, the number of hydronium and hydroxyl ions is equal and constant but adding a solute can change the concentration of hydrogen ions. Any substance that gives up or donates $H+$ ions to a solution so increasing the $H+$ content is called an acid. Substances that decrease the amount of $H+$ ions in the solution are called a base or alkali. These concentrations of acidic and base in a solution are measured in terms of the pH scale from 0 to 14 where the mid- point 7 is ordinary water which is neutral. The lower the value below 7, the more acidic is the solution. The higher the value above 7, the more base is the solution. The pH is a logarithmic scale so that a pH of five is ten times more acidic than a pH of six. Most biological processes will only occur and function properly within a fairly narrow band of pH. The variability and breadth of variability of water in this respect is therefore crucial to life. The next time you are by a lake or river or in the rain or a splash of water lands on the back of your hand, look at it and ponder for a moment. You are looking at just about the most wonderful, basic substance in our entire universe. It is utterly crucial to the assembly, configuration, integrity and function of the whole range of life systems. The Anthropic Principle derives mainly from the grand science of astrophysics but, in this respect, little or nothing could be more compelling than the extraordinary properties of ordinary water.

In this part of our brief discussion on life chemistry, as well as seeing the incredible subtlety and ingenuity of life, we see the great enigma confounding theories on its origin. Whilst Darwin's theory points to a beginning of life in much the same way as astrophysics points to a beginning of the universe, it is an aspect to which he made little reference. Unfortunately, in his time, life science was insufficiently advanced for him to even speculate. His scientific wrecking of the biblical stories led him to conclude however, that it was a natural process. The view of modern science derives of that same premise. It is interesting to note that the two greatest questions we can ask, how did the universe begin? And how did life begin? Should lead science along the same retrospective path to a barrier beyond which it seems we cannot know.

If a meteorite escapes the outer planets, enters the Earth's gravitational field and survives atmospheric friction, it will fall to the ground. In the chemistry we have thus far considered, things happen for much the same reason. They are part and parcel of the forces of which they are comprised and those which act on them. In living systems however, their course of eventuality is engineered and controlled.

Whether life derives of metaphysical or purely naturalistic influence, protein is its physical expression. Rather than things that just happen, protein does things, it is functional, it is the living stuff of everything from virus to man. It represents the visible categorical difference between the inanimate and the living. When you see a microbe squirm, an insect crawl or fly, a fish swim, an animal run or a man at work, whatever they are doing, it is protein that is doing the doing. When you look at something alive whether it is an elephant or a man or a tadpole you are looking at protein. The substances we have thus far considered, whilst being active at the molecular and atomic levels, are, to our senses, just as inert as a lump of rock or a brick wall. It is only when incorporated into the complexity of protein structure that they become alive. Some chemical reactions which, outside the sphere of life, can take a million years to react are made to react in a matter of milliseconds under the catalytic influence of special proteins called enzymes. A catalyst accelerates the speed of a chemical reaction without entering into the reaction itself. It adsorbs the reactant molecule onto its surface (Adsorb means to adhere rather than absorb which means to mix). It minimises the activation energy necessary for the reaction to occur so that the reaction occurs almost immediately and, in the process, the reactant molecule is ejected from the surface of the catalyst. The catalyst is then free to catalyse another similar reaction. A particular enzyme can manage many millions or billions of reactions before wearing out.

Modern science believes that the strange, magical influence we call life is a product of the material complexity of protein structure rather than the other way around. If you look at a nicely dressed fish in a fishmonger's window or a luscious slice of Angus steak on a butcher's slab, you are looking at materials comprised of the most complex protein structures. But they are not alive and there is no scientific course of action that can make them seem even remotely alive. They are as inert as a lump of rock. They were part of a chain, as indeed you still are, but which, for them, has been broken; a chain of eventuality which was encompassed within and sustained by the life force (By which I do not infer Vitalism but merely an inherent characteristic of life itself) and extending, complete and unbroken, back into the obscurity of unimaginable time and the very inception of life.

The science of astrophysics uses what it can see of how the universe is today together with what it was like in the past facilitated by its understanding of the electromagnetic spectrum and its relationship with time and space to work its way retrospectively to a beginning. It reaches a somewhat enigmatic and incomprehensible conclusion. In the way in which our intelligence works, you can't have nothing where there is something and you can't have something where there is nothing. In the beginning described by science there is neither something nor nothing.

Similarly, evolutionary science shows an ever-decreasing degree of both range and complexity in living organisms as both

61

the disciplines of geology and palaeontology take our understanding of Earth's history back and back into the fuzzy, dim and ultimately dark depths of geological and evolutionary time.

If we consider the age of the Earth from the point where it solidified and formed the oceans, science estimates it to be in the region of four and a half thousand million years. As of the present time, the oldest fossils, a form of cyano bacteria, go back three and a half thousand million years. Whilst these are the most primitive form of life known, they are composed of a prokaryotic cell which in itself represents a period of some tens to hundreds of millions of years of evolution. In terms of geological time spans, we therefore have to consider the beginning of life on our planet as being more or less coincident with the formation of the planet. This casts serious aspersion on any ideas of its having been a purely chance event. Of course, it may have been imported by means of comets or meteors shortly after the hadean era; the super violent, intense cooking of our planet, but it is doubtful it would have survived the method of importation. All in all, it is extremely difficult to arrive at any convincing scenario regarding when and where life began. All we can say for sure is that, on our planet, it has evolved from relatively simple organisms to what we are today over a period of some four thousand million years and, for ninety percent of that time, was confined to a single celled, microbial realm of being.

The word protein describes a wide and varied range of substances comprised largely of the molecular substances we have already considered; hair, nails, claws, foot pads, noses, bone, cartilage, muscle fibres, eyes, brain tissue, nerve fibres, cell walls, insect exoskeletons and wings, plant substances and so on. Some estimates say there are in excess of a hundred thousand different protein substances in the human body alone.

Proteins range from complex to extremely complex in both molecular and macroscopic physical structure and are, as yet, endowed with an unfathomable mystery. Most things outside the sphere of life are much more understandable in their gross form than in their fundamentals. If you take a car engine for example, almost anyone can see how it works as an engine. They can easily understand the general configuration of cylinder bores, pistons, connecting rods, crankshaft, inlet and outlet valves and their timing devices and spark distribution system. They may not be quite as aware of how the fuel is delivered to the engine. Or why the fuel works as it does or what it is made of or how the fuel to air ratio is arrived at and maintained, the differences between aspiration and fuel injection, why engines use different fuels and some have spark plugs and some don't. The more elementary aspects of engines are obscure. With proteins it is the other way around. Science is quite aware of most of the elementary aspects of protein structure; what they are made of and how and why. It is aware of all the chemical processes, the substances from which they derive and those they produce. But the gross structure of proteins is still much of a mystery. The way in which they coil, fold, roll and shape themselves is as crucial to their functionality as the substances of which they are made. Why proteins adopt their particular gross configurations and how this affects their functionality or, in other words, their aliveness, is no more understood by modern science than it was by the ancient Greeks. This, to be fair, is a bit of an over statement. Science does have a more or less complete understanding of the primary and secondary structures of proteins. Also, even though there are serious difficulties in presenting proteins to x ray diffraction techniques, it is presently making great inroads to the understanding of the more obscure tertiary and quaternary structures. But that magical and subtle difference between the fish that swims and the one in the fishmonger's window, that which makes you and I wonder, is as elusive as ever.

Proteins

I trust that until now, I have managed to keep our discussion of life chemistry brief, simple, understandable and interesting. I shall attempt to keep it so even though this is now becoming much more difficult. I would beg the reader's indulgence to bear with me however, bearing in mind the object of our discussion until now and including what immediately follows. This is to generate a conceptual frame of mind within which the reader might, with some qualification, judge for him or herself whether life began as a result of an accident or was the subject of some guiding influence.

Science is our continuing and arduous search for the greatest of beauties which are truth, knowledge and understanding. It might be considered like a river supported by many tributaries following many different courses, some of which are obscure and others more significant, but all of which are crucial to the whole. In our discussions here, I shall, in passing, mention at least a few of the people whose endeavours have led to our enlightenment and, hopefully, share with you at least a heartbeat of thanks, honour and admiration of their shining.

I mention computers here by way of some explanation but more so to highlight the magnificence, brilliance and ingenuity of life. We are all familiar with the tremendous facility the modern computer provides. All of it is based on a simple code called binary which is comprised of noughts' and ones. These 0s and 1s do not represent arithmetic or any other values. They represent two different states. These two different states are physically engineered on the computer by switches. The two different states of a switch are ON and OFF. Computer programmers use this code in conjunction with its application to various semi-conductor gates and mixes, where it is expressed as a square voltage waveform, to construct more than three hundred different languages of which our computer facility is comprised.

Life uses a much more elaborate system of coding and language construction. You may recall, when we mentioned Pyridine, that a new element entered our discussion i.e. Nitrogen. Nitrogen plays a crucial role in both the coding and construction of the language of life. You may also recall that the COOH group confers acidic properties on any molecule of which it is part. Nitrogen and two hydrogens NH_2, derived from ammonia NH_3, does the opposite. It confers the properties of a base or alkali on any molecule of which it is part. Thus, we have a differentiation of two states, acid and base.

There is only one kind of computer switch, one kind of on and off. Also, the computer is a virtual world as opposed to life which is a reality. Life's coding must also be a reality therefore and, rather than just representing information, must also be compatible in terms of physical reactivity with the chemistry of which it is part. So, although life has at its disposal the same differentiated state method of coding as that used in computers, it does not in fact use that method. Instead, it uses a sequencing method expressed through five different nitrogenous bases supported on an acidic sugar phosphate framework the whole of which is referred to as the nucleic acids. The language of life is therefore significantly more comprehensive than that of computers. Further to this, the vocabulary of life's language is almost infinitely broader than that of any other language. In our language for example, we have twenty-six characters together with a few punctuation signals. Our longest words in normal usage are comprised of some twenty characters and no character is repeated more than twice. In the vocabulary of life there are twenty characters called the amino acids and words can be many thousands of characters long and can contain long chains of repeated characters. It is the most extensive language on Earth; it is coded for in the nucleic acids and dictates the structure, characteristics and functionality of proteins. It dictates what you are from your toenails to how you interact with those around you. A protein is comprised essentially of numbers and assortments of different amino acids joined together in a polypeptide chain and coiled, rolled or folded into its final functional shape. To be more accurate, the chain is comprised of amino acid residues since a molecule of water is lost from each in forming the peptide bond.

The Nucleic Acids

In any consideration of what life is, the nucleic acids have to be the starting point. They represent the most fundamental aspect of life and the question of how they came to be is one of the most intriguing in the whole of science. They are manufactured and, for the most part, reside within the nucleus of the cell. See Fig. 9.

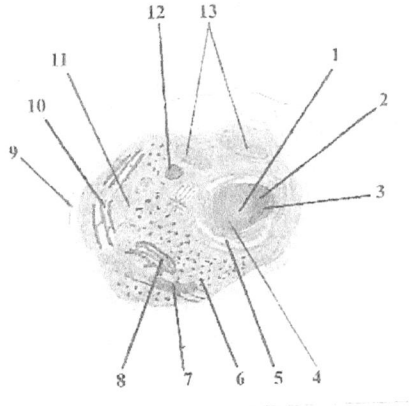

1 Nucleolus, 2 Nucleus, 3 Nuclear pores, 4 Chromatin, 5 Rough endoplasmic

reticulum, 6 Ribosomes, 7 Smooth endoplasmic reticulum, 8 Golgi apparatus,

9 Plasma membrane, 10 Smooth endoplasmic reticulum, 11 Cytoplasm, 12 Lysosome,

13 Mitochondria.

Fig. 9 Eukaryotic Cell

Whilst up until now, I have spoken figuratively of the aliveness of proteins because they represent the functional substance of all living things, in fact, the smallest entity that can be said to be alive is the cell. Proteins are manufactured in the cell and the average adult human body is comprised of some one hundred trillion cells which, in themselves, are comprised essentially of protein. Just as your body has organs such as your brain, heart, lungs, liver, kidneys etc. to carry out special functions, so the cell also has minute, specific purpose functionaries. But in the cell, they are called organelles. Some of these are shown in Fig. 9.

In the nucleic acids, we encounter another new element, phosphorus. Phosphorus is a dangerous and highly poisonous substance. It is of the nitrogen group in the periodic table and has some similarities with nitrogen

66

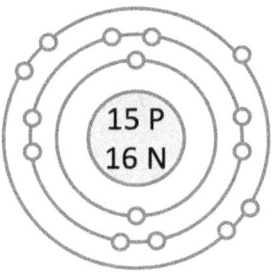

Phosphorus

You will remember that nitrogen has two orbitals and has five electrons in its outermost. Phosphorus has three orbitals and also has five electrons in its outermost. It is a highly reactive element, so much so, that it is rarely if ever found in its elementary form. It is always in a molecular form bound with something else. It is absorbed from the soil by plants and in turn by animals that eat plants in the non-poisonous form of what is called a phosphate where it is in molecular combination with oxygen. Here again we see, as in table salt, that a change in electron arrangement (Or force field configuration) resulting from molecular bonding can render a deadly poisonous substance amenable to life.

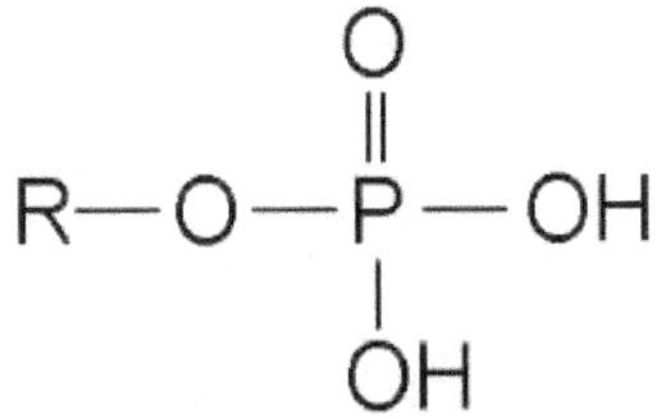

$$R-O-\overset{\displaystyle O}{\overset{\displaystyle \|}{\underset{\displaystyle |}{\underset{\displaystyle OH}{P}}}}-OH$$

[OBJ] Phosphate group

(When two or more atoms are bound together forming part of a molecule and have a distinct effect on that molecule's overall function, they are often referred to as a group such as the "phosphate group" and the "COOH" group" mentioned earlier. The R in the diagram above indicates its connection to the rest of the molecule).

We also encounter two new sugars, ribose and deoxyribose made mainly in animals and obtainable in small amounts from ripe fruits. They are both pentoses, five carbon sugars, and the only difference between the two is that the OH hydroxyl group attached to the second carbon in ribose is replaced by a single hydrogen atom in deoxyribose. Alternatively, the difference between the two is that deoxyribose has one oxygen atom missing, hence its name. (A hydroxyl group is an oxygen atom joined by a single bond to a hydrogen atom).

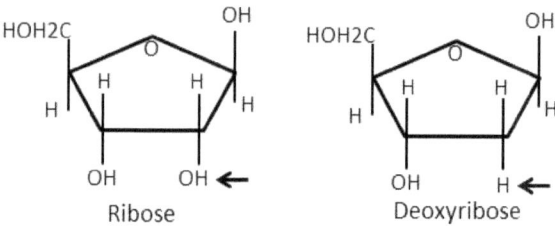

Ribose   Deoxyribose

[OBJ]If one were describing the building of a house or anything else, there would be a distinct starting and finishing point; not so with the nucleic acids. The nucleic acids in your body, and in everything else that lives, are part of a continuous, ongoing chain of living activity that stretches back some three and a half to four thousand million years into Earth's past. This is the fascinating intrigue with which our minds shall wrestle shortly when we consider how they might have come to be.

The nucleic acids are complex molecules which are called nucleic acids because they are assembled mainly within the nucleus of the cell. They are comprised of two main classes named after the names of the sugars which make up what is called their backbone, ribose and deoxyribose. However, there is nothing rigid or boney about this part of the molecule but, as well as being functional, it does provide the molecule with structural configuration and integrity. So, the two nucleic acids are called ribonucleic acid or RNA for short and deoxyribonucleic acid or the now famous DNA, the existence of which you will already be familiar with. The RNAs are derived from DNA. Of all the characteristics descriptive of and attributable to life, replication, reproduction, and heredity are its most powerful and fundamental and are the remit of the nucleic acids.

Whilst war horses, farm animals and crops and domestic animals had been the subjects of selective breeding for some thousands of years, it was, until the first half of the twentieth century, a rather hit and miss affair. However, in the eighteen sixties, a rather obscure monk, Gregor Mendel, teaching mathematics and physics in a monastery in Brno (Now in the Czech Republic), succeeded, over a decade of disciplined scientific research on plants, in extrapolating the rules of heredity. He did this at a time when there was no knowledge whatsoever of the existence of nucleic acids. His work was so far ahead of his time that it lay unrecognised for best part of half a century. It was rediscovered in nineteen hundred, long after his death, by the Dutch botanist Hugo d'Vries together with other contemporaries who used it to verify their own independent findings. This is when the science of genetics is said to have been born. Mendel's rules still stand, and he is now sometimes called "The Father of Genetics".

The nucleic acids were first discovered, though not identified as such, in the eighteen eighties by the Swiss biologist J. Friedrich Miescher, then in the University of Tubingen in Germany. He called them nuclein. They have been the subject of much arduous research

Gregor Mendel Courtesy Wikimedia commons

ever since. Miescher also succeeded, some time later, in breaking his nuclein down into part protein and part nucleic acids. During the nineteen twenties, they were associated with chromosomes, the gene packaging units in the nuclei of cells. However, at that time, it was thought that the nucleic acids were too simple a conglomerate to accommodate the storage capacity for replication and the enormous amount of information representing heredity. It was thought that all of this must be contained within the more complex structures of protein. A lack of sulphur in the nucleic acids was noted as a significant difference from the molecular make-up of proteins which contain a significant amount. Also, the nucleic acids contain a lot of phosphorus, of which proteins contain only a small amount. These differences later turned out to be crucial in finally identifying the nucleic acids as life's information storage complex. It is interesting to note at this point, that whilst there are those who question the validity of research projects that do not offer immediately foreseeable benefits, benefit from science for its own sake is as inevitable as day following night.

Courtesy Wikimedia commons
Johannes Friedrich Miescher

Whilst the rules of heredity had been established in the eighteen sixties and the nucleic acids discovered in the eighteen eighties, by as late as nineteen fifty, the locus of life's informative compendium was still unknown. However, in nineteen fifteen a new window which was to throw a most significant light on this understanding emerged from a quite unrelated course of research.

In eighteen fifty-five, a lawyer called Thomas Brown M.A., LL.B., left a legacy to the University of London for the purpose of setting up a charitable veterinary practice and research facility. This was finally established in Wandsworth Road in eighteen seventy-one. It was called The Brown Animal Sanitary Institution. It proved to be a centre of excellence in terms of its primary purpose. Also, because of the particular leanings of its first professor-superintendent, Sir John Burdon Sanderson, who contributed financially to its establishment, it also contained the first comparative pathology research centre to be set up in the United Kingdom. It became home to some of the most distinguished

names in the fields of comparative medicine and physiology including Charles Scott Sherrington, a Nobel Prize winner in the field of neuroscience. Unfortunately, the institute was rather short lived. Its primary objective underwent gradual dissolution due mainly to the replacement of horses by steam and internal combustion engines and it was bombed out of existence in nineteen forty-four by a flying bomb.

Dr. Frederick Twort

We are concerned here with The Brown Institute's final and longest sitting superintendent, Dr. Frederick Twort. In his first ten years of tenure, he made several significant contributions to medical science and the alleviation of some serious cattle diseases. However, he was a man who did not sit well with his financial and administrative superiors at the University of London. He was want to venture into realms of research that did not meet their financial approval. In course of this, he was the first person to discover, in nineteen fifteen, that infectious bacteria were themselves subject to infection by other incredibly small agents; he called them bacteriolytic agents, (Viruses). Two years later, the same discovery was made independently by the French-Canadian microbiologist Felix d'Herelle, who called them by their present name, bacteriophage. However, I should mention that Dr. Twort had published a paper on the subject by that time and, whilst d'Herelle received the accolade for the discovery, he was not considered a man of great integrity by the scientific community at that time. It would take another three and a half decades before someone would recognise the value of this most crucial discovery in terms of its applicability to the discovery of where and how life secreted its three and a half billion-year-old library of replicative and hereditary information.

In the mid nineteen thirties, the Russian scientist Andrei Nikolaevitch Belozersky isolated clean, pure DNA for the first time. Also, during that period, and in fact, from the early nineteen twenties, methods for crystallizing biological substances together with development of sophisticated instrumentation and associated mathematics, especially at the Carnegie Institution of Washington Geophysical Laboratory, were establishing X ray crystallography as an applied science. This is a means of determining the arrangement of atoms within the crystallised substance. In the mid nineteen thirties, this branch of science was also being pioneered by the eminent British Irish scientist John Desmond Bernal. I note here that two of the people studying with Bernal at that time were Dr. Rosalind Franklin and Max Perutz. These names will appear again shortly and will demonstrate that the scientific community, our intellectual elite, is no more immune to back-stabbing and self-presentation for accolade and promotion over one's colleagues than any commercial laundry or local authority waste collection department.

In the mid nineteen twenties, the British scientist, Frederick Griffith, demonstrated that some bacteria exist in two forms, in the case of his particular experiment, pneumococci. Pneumococci was found to exist in two forms, one smooth and one rough and named

72

by Griffith S and R. The smooth form's smoothness results from its

encapsulation within a capsule that resists attack by the body's immune system.

Frederick Griffith
Courtesy Wikimedia commons.

It is therefore very virulent. The R form has its cell wall as its outer surface and is non-virulent due to its vulnerability to attack by the immune system. More importantly, Griffith discovered that, during a heavy or concentrated infection, the R form could, in some instances, acquire a capsule and become the deadly S form. He also discovered, during experiments with mice, that if he injected dead S form, the host was unaffected. But if he injected the dead S form together with live R form, the host always died. By extracting the pneumococci from the dead host he discovered that the R form had changed into the virulent S form very readily in the presence of the

dead S form. He had discovered bacterial transformation. But the means whereby R was able to achieve this apparently objective change in form and function was unknown.

What had actually happened here was that, for the purpose of his experiment, Griffith had killed the S form of his bacteria by boiling. But whilst the S form was indeed dead,

Oswald T. Avery Courtesy Wikimedia commons.

its DNA component had survived the boiling and was able to make its way into the living R form; so, transforming it into the deadly S form.

During the mid- nineteen forties, the American researcher, Oswald Avery continued and extended Griffith's line of research. By this time, better methods were available for breaking down biological substances. Avery first removed the larger cellular structures from S strain bacteria. The bacteria were then treated with enzymes which removed all the protein from the cells. What was left was then mixed with R strain bacteria; The R strain still transformed into the S strain. These experiments started a serious scientific leaning towards the idea that the nucleic acids rather than the more complex protein were the carriers of genetic information.

The nucleic acids became the subjects of wide and intensive research. Their general chemical make-up in terms of atoms and molecules were well known to researchers but this did little or nothing to explain how the replication and hereditary information was stored. They knew, for example, that DNA was comprised of single molecules of the sugar deoxyribose interspaced by a phosphate group and that the sugar molecules were connected to groups of purines and pyrimidines comprised of adenine, thymine, guanine and cytosine. (A purine is a heterocyclic, aromatic, nitrogen containing organic compound). In the cases of adenine and guanine they are bicyclic, meaning they each have two carbon rings. The pyrimidines are monocyclic having only one carbon ring). However, most researchers believed that the DNA molecule contained a constantly repeating sequence of G, A, C and T which made little sense of any instructive code.

At the end of the nineteen forties, the American biochemist Erwin Chargaff proposed two rules of extreme significance, now known as Chargaff's Rules. Using the newly developed techniques of paper chromatography and the ultraviolet spectrophotometer, He showed that in DNA the number of guanine units equals the number of cytosine units and the number of adenine units equals

the number of thymine units and that variations in the numbers of combinations existed. This led to his second rule proposing that the composition of DNA varies from one species to another in terms of the relative amounts of A, G, T and C bases. This molecular diversity in DNA made it much more credible as a possible information storage complex.

Erwin Chargaff.

In nineteen fifty-two, the American biologists, Alfred Hershey and Martha Chase, carried out a most crucial and what is described nowadays as a simple experiment. It may have been simple in terms of logic and methodology but it would not have been possible were it not for the work and hard gained understanding achieved by Twort and d'Herelle, Griffith, Levine, Chargaff and Avery. It also made use of the already known chemical compositions of protein and the nucleic acids especially in terms of their sulphur and phosphorus contents. Nevertheless, it was a brilliantly conceived experiment, made possible by collating this knowledge and applying the relatively new technique of radioactive marking. Hershey and Chase also knew that viruses were a relatively simple organism comprised only of nucleic acids and protein and that they pro-created by invading cells of their hosts and inducing them to stop replicating themselves and to create new viruses instead. Viruses had the capacity to monopolise the replicative machinery of the host cell. Hershey and Chase reasoned that, if they could determine which part of the virus entered the host's cell replication machinery, they would, at last, know where life's coded instructions lay.

Hershey and Chase cultured several batches of bacteria including one in a medium containing radioactive sulphur and another in a medium containing radioactive phosphorus. They infected both of the radioactively marked bacteria cultures with viruses and succeeded in culturing two new lots of viruses, one marked with radioactive sulphur in their protein and one marked with radioactive phosphorus in their DNA. They then infected one lot of normal bacteria with the viruses containing radioactive sulphur and another with the viruses containing radioactive phosphorus. These viruses induced their respective hosts to produce more viruses. The two separate lots of new viruses were then subjected to various cleaning and separating processes. Upon detailed examination, it was found, to the delight of Hershey and Chase that the bacteria culture infected by radioactive sulphur produced normal, unmarked viruses and the culture infected by radioactive phosphorus produced radioactive marked viruses. This was proof that the DNA component of the virus had entered the host's cell replication machinery and the protein component had not. As a final proof, Hershey and Chase infected another lot of normal bacteria with this newly produced, marked virus and found that the further progeny of the virus was similarly marked. Thus, it was finally established that the instructions for building living things were contained in DNA. The question now, the most crucial question across the entire spectrum of human scientific endeavour to date, was HOW?

The answer to this question became the Holy Grail of the scientific community at the start of the nineteen fifties. Whilst the structure of DNA is considered simple in comparison to that of protein, it is still a highly complex molecule. This is evidenced by the fact that one of the most eminent scientists of the time, Linus Pauling, and the Russian American biochemist Phoebus Levine, who had discovered the molecular make-up of DNA, proposed different secondary structures of the molecule both of which were wrong. However, it was still thought by many that Linus Pauling, who was at the cutting edge of understanding biological molecules and who had already elucidated complex protein structures, would be first to discover the secondary structure of DNA.

Amongst the many laboratories struggling with this question, were two in the United Kingdom operating under the auspices of the Medical Research Council. One was the Cavendish Laboratory at Cambridge University headed by Sir Lawrence Bragg where the English physicist and biologist Francis Crick and the American zoologist and biologist James D. Watson were attacking the problem. The other laboratory was at King's College in the University of London where the DNA molecule was being researched by a team who were also further developing the science of X- ray crystallographic and fibre diffraction imagery and interpretation. This team, under the directorship of Sir John T. Randall, included the British biophysicist and chemist Dr. Rosalind Franklin, the New Zealand-born physicist and molecular biologist Maurice Wilkins, the physicist Alexander R. Stokes and the physicist Herbert R. Wilson.

At the Cavendish laboratory, Watson and Crick had meticulously collated and interpolated all that was known about the DNA molecule to date and had brilliantly conceived of the idea that some kind of code might be embedded in base pairing of the differentiated states of the purines and pyrimidines. On this basis, they set out to build a ball and stick physical model of the molecule. However, their every attempt was being agonisingly frustrated by one small error in their idea of how the molecule was put together. They were unable to achieve a regularly repeating conformation of the molecule. Their error was a rather technical one involving a phenomenon known as tautorism. Guanine and thymine are tautomers. A tautomer is a molecule that can spontaneously and intermittently change its three-dimensional shape because of a hydrogen atom that flips from one position to another within the molecule. The two different configurations arising from this are called enol and keto. Watson and Crick had persisted in using the enol form in their model which was the wrong shape. The keto form is the one which is used and stabilised within the DNA molecule.

At the King's College laboratory in the University of London, Dr. Rosalind Frankiln and Maurice Wilkins were examining the DNA molecule in a much more technical manner. They were experimenting with the molecule in various states of hydration and photographing it by means of X-ray crystallography and fibre diffraction then applying the associated mathematics to interpret the results. The DNA molecule exists in two forms of equal amounts, the A and B forms. This makes X-ray diffraction results ambiguous. Dr. Franklin, using highly skilled chemistry techniques, succeeded in separating the two forms and eventually produced the famous photo no. 51 of the B form of DNA. On

Courtesy Wikimedia commons
James D. Watson   **Francis Crick**

interpretation, this showed the molecule to be double stranded with the backbones on the outside, interconnected by base pairs and the whole twisted into a double helix form of consistent diameter.

In nineteen fifty-two, Dr. Franklin sent a progress report of her work to the Medical Research Council at Cambridge. It somehow came into the hands of her former student colleague, Dr. Max Perutz, who was researching haemoglobin there at that time.

Without her knowledge or permission, Perutz showed Dr. Franklin's report to Watson and Crick. Watson could make little or nothing of it since he wasn't qualified in X-ray crystallography. Crick, on the other hand, who was a physicist with experience of the technique and the associated Fourier transforms of Bessel functions representing the X- ray diffraction patterns of helical structures of atoms was immediately able to recognise from it where their frustrating little error lay. He was also able to further deduce from the text of Dr. Franklin's findings that DNA was a double helix with the two polynucleotide chains running in opposite directions. Watson and Crick were now able to successfully build their ball and stick model of the secondary structure of the DNA molecule. They published their results with little mention of Rosalind Franklin's crucial part and received all the accolade for the discovery even though Rosalind Franklin had beat them to it.

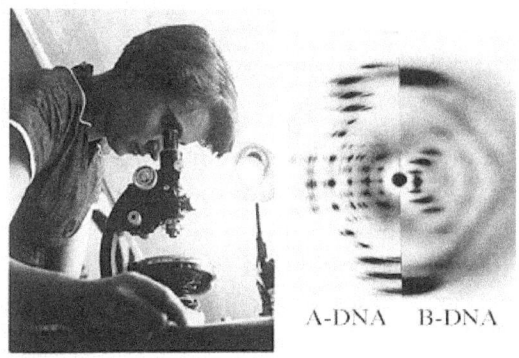

A-DNA   B-DNA

Dr. Rosalind Franklin and her famous photo 51

Courtesy Wikimedia commons

Dr. Rosalind Franklin was a woman, of which there were few in the scientific community in her time. She was also the daughter of a Jewish, London banker. She could have spent her time playing tennis, yachting and socialising but she was driven by the dynamic enthusiasm and insatiable passion for truth and understanding of nature that drives all eminent scientists. She should be remembered for the crucial part she played in our understanding of the fascinating workings of life.

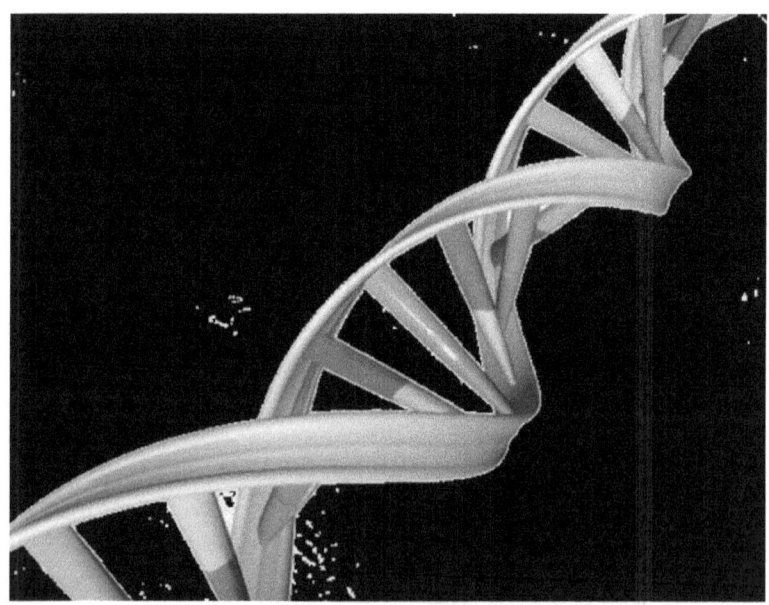

DNA Helix: courtesy Wikimedia commons
Fig. 10. Typical representation of the DNA double helix
.
Of course, there is no doubt that Watson and Crick or possibly Linus Pauling or some other in the field of genetics at that time would have discovered the wonderful secret of DNA. The accrued knowledge and climate of thinking were right for it at the time and only a structural error was frustrating Watson and Crick. But Dr. Rosalind Franklin did discover it first. Immediately following the revealed understanding of this aspect of the DNA molecule, the science of genetics underwent explosive growth but Dr. Franklin was only able to enjoy a short part of it. She was ill for most of the five years immediately following her discovery. Even so, she did succeed in that time, in extrapolating the workings of the strange

tobacco mosaic virus. She finally died in nineteen fifty-eight, aged only thirty-seven. Many believe that, like Marie Curie, she paid the price for our understanding with her life. She died of ovarian and abdominal cancer, probably caused by her extensive use of powerful X-ray machines.

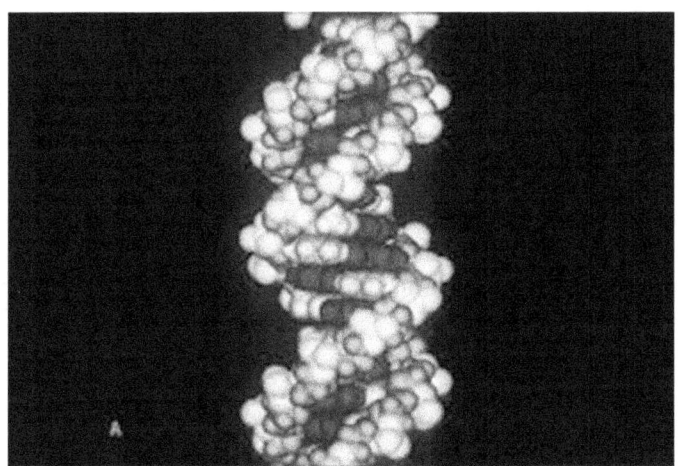

Courtesy Wikimedia commons
Fig 11 The DNA molecule.

Fig. 10 is a schematic representation of the DNA molecule illustrating the kind of base pairing arrangement that forms its information content. Fig. 11 is a more realistic image but if you were to handle it, it would in fact be just a sliver of snot. Our understanding of the DNA molecule is the result of hard and dedicated science together with sophisticated calculation rather than by observation. There is no limit to the praise and honour due to those who devoted large parts of their lives to extrapolating the overwhelming beauty and apparent ingenuity of this most fundamental aspect of the strange and intriguing phenomenon we call life.

Notwithstanding the fracas of accolade, doctors Crick and Watson were crucial participants in the early science of molecular biology. In course of building their physical model of the DNA molecule, from Dr. Franklin's findings, they had assimilated and understood all that was known of it to date. As a result, they were able to postulate, from their model, the most fundamental, the most fascinating and intriguing trick of life; that which supports both its persistence and evolution. They postulated the mechanism of replication. It was mentioned, only somewhat tangently, at the end of their 1953 published article on the DNA molecule in their famous statement "It has not escaped our notice that the specific pairing we have postulated immediately suggests a possible copying mechanism for the genetic material". Following publication, Crick and Watson strutted proudly into their club-room and announced to their friends and colleagues "We have discovered the secret of life". Great as their discovery was, little did they know at that time that they had prised open but a tiny chink in a doorway leading to a micro-universe of mind boggling yet wondrous complexity.

Having some understanding of what the DNA molecule is as an information store, we shall now have a look at what science has learned of the biological flow of the information and its expression in the physicality of our being. This occurs in three main stages called Replication, Transcription and Translation.

Chapter 4 - The Central Dogma

We shall have a superficial glance at this wonder of wonders; this which Crick and Watson came to call "The central dogma of molecular biology" and for which they, together with Maurice Wilkins (Dr. Rosalind Franklin 's supervisor) jointly received the Nobel Prize in 1962. If the wonders of our universe were to be numbered in accordance with their greatness, number one would be intelligence followed immediately by the DNA molecule. Not in terms of chemical complexity, vastly complex as it is, but in its encompassment and perpetuation of that strangest of all forces in our universe, life.

DNA exists in two designated organelles within the cell. One is the mitochondrion of which there are a large number distributed throughout the cytoplasm. Even so, mitochondrial DNA represents only a small fraction of the total, and, interestingly, it does not conform with the standard genetic code. It has the capacity to replicate within the cell and contains thirty-seven genes some of which carry the instructions for making enzymes and others instructions for making transfer and ribosomal RNAs. Mitochondrial DNA has the peculiarity that it is inherited only from the mother of progeny. It is useful in forensics and can be used to trace maternal heritage over many generations. It has other extremely interesting aspects which we shall briefly consider later.

The main bulk of DNA in humans is packed into forty-six chromosomes in twenty-three pairs and confined within the nucleus of the cell. It carries the totality of information representing our being and is the instrument for cell replication.

When an architect designs a building or an engineer designs an intricate machine or a computer programmer designs a game or operating system, all of these things are found, upon completion, to be beset with faults or bugs. De-bugging therefore becomes an essential part of the overall process. As we shall see, even in our very superficial look at the replication process, nature has apparently encountered similar and similarly dealt with this aspect of intricate functionality. You might note it in the action of enzymes and the high degree of functional specificity to which they are attuned. Irrespective of your particular orientation regarding the God idea, I should be surprised indeed if you do not wonder, at least a little, about the secular approach of scientists who have a much deeper understanding of the intricacies of this spectacular aspect of life.

Replication

Again, this is not a chemistry lesson. You do not need to fully understand the detailed chemistry of the DNA molecule nor indeed the jargon associated with it in order to appreciate the wonder and the beauty of it. What we have already covered will suffice for this. I shall merely add that you will notice some numbering which makes some directional reference in some parts of the DNA molecule. This comes from the international I.U.P.A.C. system of nomenclature that scientists use, amongst other things, to differentiate between atoms in a molecule. It would require extensive explanation which we shall avoid here as you will easily understand its implications, in terms of our discussion, as we go. Also, you should be prepared for some confusion insomuch as we are dealing with a circular, interdependent mechanism, like the chicken and egg question, where we are breaking into the circle only at points relative to our particular severely abridged

84

discussion. For example, DNA is manufactured in the nucleus of the cell but the cell is comprised essentially of protein. Protein is manufactured under the directorship of the RNAs which themselves are copied from DNA. Do not be concerned by this confusion; it is the great enigma of life to which this part of our discussion inevitably leads.

Starting with the sugar phosphate backbone of the DNA molecule, you will see, from Fig. 10, that there are two of these running parallel and twisted into a helical shape. Each is comprised of an alternating chain of phosphates and sugars. In the case of DNA, the sugar is deoxyribose. You can easily visualise that, if these were straightened out, you would have two straight, parallel sides interconnected by the base pairs and looking very much like an ordinary ladder. You might imagine it looking something like Fig. 12.

Fig. 12. Sketch of DNA ladder

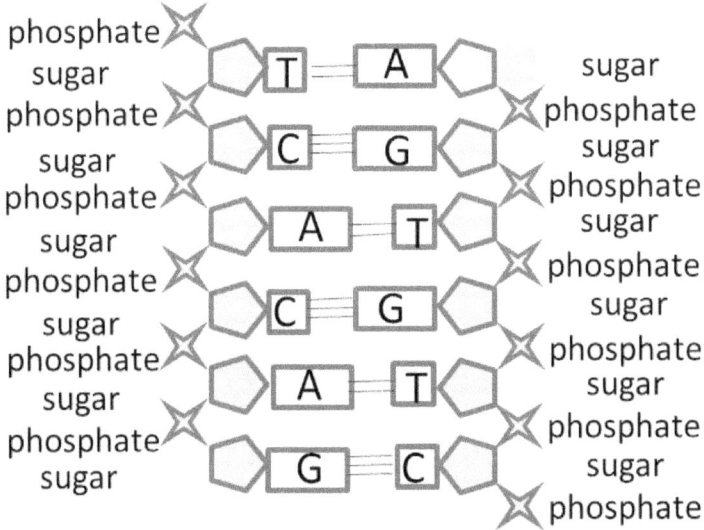

The first thing you might notice in Fig. 12 is that C - Cytosine always pairs with G - Guanine and T - Thymine always pairs with

A - Adenine. This is in perfect compliance with Chargaff's rule which he proposed before the makeup of the molecule was known. You will remember he stated that the amount of cytosine equalled the amount of guanine and the amount of thymine equalled the amount of adenine. The method students generally use to remember which goes with which is, that the capital letters starting the names Cytosine and Guanine, which go together, are both comprised essentially of curved lines and those starting Thymine and Adenine, which go together, are comprised of straight intersecting lines. As a matter of trivial interest but more importantly of perspective, the length across any single base pair is 1.085 nanometres. This means it would take some sixty-five thousand of them placed end to end to stretch across the diameter of a single human hair. The DNA molecule is about two meters long and comprises some three billion base pairs.

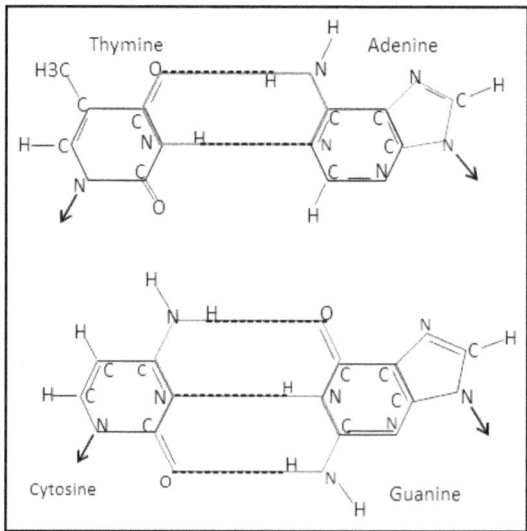

Fig. 13 Base pairing

Fig. 13 shows two base pairs in the order TA and CG. These also occur in the order AT and GC as you can see in Figs. 10 and

12. Adenine, thymine, guanine and cytosine are bases but their base properties are not important in the coding of the information. The information is carried in the differentiated states of the purines and pyrimidines of which the bases are comprised and the sequences in which these occur, like the noughts and ones in your computer, except that the bases are active rather than passive. In Fig. 12, you will see the short lines representing the hydrogen bonds between the bases, three for the CGs and two for the TAs. Now imagine you were to cut these all the way down the centre of the molecule and pull the two strands apart. Because the bases can only pair in a specific manner, which is complimentary, if you now take an assorted bunch of new nucleotides and add them one at a time to their opposite numbers on the two single strands, you will end up with two double stranded DNA molecules exactly the same as the one you started with. The cell actually does this. It is called Replication and is what Crick and Watson were referring to in their famous statement at the end of their 1953 published article. When a base bonds to a sugar, deoxyribose in the case of DNA, you have what is called a nucleoside. When the sugar further bonds to a phosphate by means of a phosphodiester bond you have a nucleotide. The DNA molecule is comprised of two chains of nucleotides bonded together by hydrogen bonds between the bases. As you can see in Figs. 12 and 13, thymine and adenine bond with two hydrogen bonds and cytosine and guanine bond with three hydrogen bonds. The cytosine- guanine bonds therefore assist in stabilising the molecule. A nucleotide is a unit or monomer of the chain and the chain is formed by the phosphates further bonding to the sugars so connecting the units together. This polymerisation forming the chain is directional.

Imagine you are washing your car in the street with a pressure washer. In order to reach the nearest power supply socket, you need to use two electrical extensions. You know intuitively that these must lie in a certain order. It's no use if you have them in a plug to plug or socket to socket order. They must lie in the order, supply socket to extension plug, extension socket to extension plug, extension socket to pressure washer plug. The nucleotides in a strand of DNA have a similar arrangement. The sugars are pentoses, five carbon sugars, and the carbon atoms in the molecule are numbered 1' to 5'. The primes beside these numbers are to distinguish the sugars from other parts of the molecule. The overall molecular arrangement of a nucleotide determines that the phosphates joining the sugars together in the chain can only react and bond with the 3' and 5' carbons of the sugars. This in conjunction with the chemistry of the cell machinery that facilitates this bonding determines that it can only occur in the 5' to 3' direction as shown top to bottom in Fig. 14.

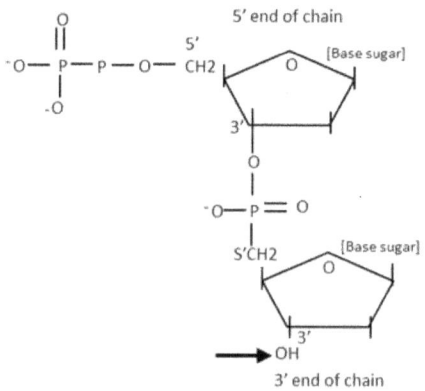

Phosphodiester bonding

Fig. 14

Note, (Arrowed) the free OH group attached to the 3' carbon at the bottom of Fig. 14. This is the socket into which the phosphate of the next nucleotide will plug in order to continue the polymerisation of the chain. A chain is measured by the number of bases and may be many thousands or tens of thousands of bases long.

Looking again at Fig. 12, you will see that the left- hand chain of nucleotides starts with a phosphate at the top and ends with a sugar at the bottom. This chain is running in a 5' to 3' direction top to bottom. The right- hand chain is starting with a sugar at the top and ending with a phosphate at the bottom. This chain is running in a 3' to 5' direction top to bottom. In this respect, the two chains of the DNA molecule are said to be anti-parallel.

A phosphodiester bond, so far as our discussion is concerned, simply means two sugars with a phosphate in between. However, in order to affect these bonds a certain amount of energy is required. You will remember that a base bonded to a sugar comprises a nucleoside and when the sugar is further bonded to a phosphate it becomes a nucleotide. The cell machinery actually produces large numbers of bases bonded to sugars which are further bonded to three phosphates. At this stage, they are called nucleoside triphosphate. The mitochondrion, in conjunction with other cell components produces eight different kinds of these which in DNA are:

dATP - deoxyribose adenosine triphosphate.
dTTP - deoxyribose thymine triphosphate.
dCTP - deoxyribose cytosine triphosphate.
dGTP - deoxyribose guanine triphosphate.
In RNA they are:
rATP - ribose adenosine triphosphate.
rTTP - ribose thymine triphosphate.
rCTP - ribose cytosine triphosphate.
rGTP - ribose guanine triphosphate.

Both the DNA and RNA nucleotide chains grow by means of a condensation reaction which removes the two extra phosphates and in doing so provides the energy necessary to bond the remaining phosphate to the next sugar in the chain. So, the DNA molecule is comprised of four sugar-base nucleosides which are phosphorylated into nucleotides, hydrogen bonded into complimentary pairs and polymerised in a 5' to 3' direction into long chains by means of phosphodiester bonds.

Living things grow and are maintained through the seeming miraculous capacity of the cell, under the directorship of the nucleic acids, to create within itself a duplicate of itself and then to divide into two identical cells. During this process, many sister cells die, are fragmented and their component parts recycled in the manufacture of new cells. This is going on at this instant, as you read, in many of the hundred or so trillion cells that make up your body. If you read for an hour, a billion cells will have been replaced in your body in that time. Even more amazingly, this is the culminate evolved state of a continuous, ongoing, unbroken, undying chain of immortal, living activity that has been passing from generation to generation for the past three and a half thousand million years. It's a mind-bending concept.

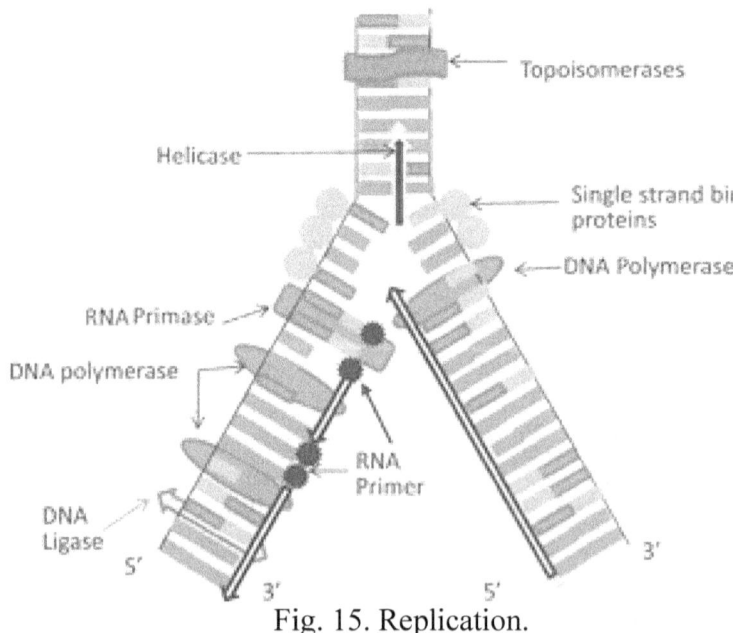

Fig. 15. Replication.

Fig. 15 shows the DNA molecule unwound from its spiral shape into the ladder form and in course of being split apart and replicated in the way previously described. In reality, Fig. 15 would not be in this straight-line geometrical shape. It would be more of an oval bubble shape.

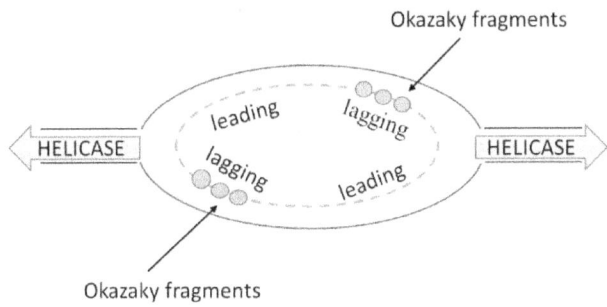

Fig. 16. Replication Bubble.

Tremendous strides in understanding have been made since Dr. Franklin's time. But such is its complexity that even now, more than half a century after Crick and Watson's revelation, there are aspects of the DNA molecule's workings that are not fully understood by science. It is still a subject of intense research. Even so, science has to be greatly admired for what it has learned and understood to the extent of having the capacity to study the life principle itself in both in vivo (Living) and in vitro (Dead) processes. From this have emerged new fields of subtle diagnostics together with vast industries manufacturing crucial therapies and treatments which would otherwise have been unimaginable. However, some of the enzymes to be mentioned are more clearly identified and understood in bacterial DNA than they are in eukaryotic DNA. Since the general chemical make-up, specification, purpose and function has to be very similar in both cases, I shall not protract by differentiating between them. It is not necessary since our purpose here is merely to generate a lay-appreciation of the wonder, subtlety, apparent ingenuity and intricate complexity of the replication process.

As you can see in Figs. 15 and 16, the breaking of the hydrogen bonds between the bases, (Just like cutting down the centre of Fig.12) results in two opening Forks emanating from a single Point of Origin and moving in opposite directions along the DNA molecule as further splitting continues. Fig. 15 is one of these forks and it is to Fig. 15 that we shall now refer.

(You should note that operations described by words like splitting, unwinding, cutting, re-joining, adding, disentangling, etc. are not carried out with axes, hammers, nails and saws. They all involve sophisticated chemical reactions catalysed by specific enzymes and other associated proteins occurring in response to subtle, sequentially timed, interrelated and closely co-ordinated chemical signalling mechanisms).

Just as our bodies go through a number of stages or phases in our lifetime, so the cell also goes through a number of phases. Replication occurs during what is called the S phase, (Synthesis phase). The information representing your entire being is contained within the nucleus of the cell. It is comprised of six billion nitrogenous bases in three billion base pairs coded in the language of life and packed into forty-six chromosomes. This, in total, comprises the DNA molecule; an informative representation of your entire being. The code itself is in the form of three-bit words; groups of three nucleotides, called codons each of which specifies a particular amino acid. The amino acid in turn will form part of a polypeptide chain of amino acids forming a protein. The entire recipe for the protein together with the instructions for its assembly is contained within a wider stretch of the DNA molecule called a gene.

The cell generates an environment within which the complex replication machinery called the Replisome is assembled. This takes the form of a bubble, as seen through a microscope. A considerable number of these bubbles form along the length of the chromosome which, in eukaryotes, is linear as opposed to bacteria where it is circular. The bubbles form on sections of the DNA molecule which are rich in AT base pairings. This action appears to

92

pre-suppose that AT base pairs, being bonded together by only two hydrogen bonds, are easier to break apart than the CG pairs which have three hydrogen bonds, even though every base pair has to be split apart. This starting point is called the Point of Origin, and, as replication proceeds, the bubble lengthens along the molecule, in both directions, until it meets the bubbles extending from adjacent sites of origin. The following steps in replication occur, by action of the replisome, within the environment of the bubble.

The packaging of the DNA molecule within the chromosome is in the form of a coil wound on a spindle; although there is no actual spindle. This begins to unwind into the helix form of the molecule. Initiator proteins attract the enzyme Helicase which starts to unwind the double helix into the straight ladder form, as in Fig. 12, and to break the hydrogen bonds between the base pairs. This causes further twisting called supercoiling and generates stresses on the molecule ahead of helicase. If allowed to continue, this would cause serious damage to the DNA molecule. These stresses are relieved through the action of the enzyme Topoisomerase 1 which cuts a single sugar-phosphate backbone so allowing the two strands to swivel around each other. This relieves stresses but tends to cause some entanglement and even knotting of the strands. Topoisomerase 2 intervenes and cuts both strands and works to disentangle and even un-knot them. The action of topoisomerases is only one of many fascinating aspects of replication. Upon being cut in the super-coiled condition, the DNA strands spin through one another at incredible speed. The topoisomerases go with them and are charged with affecting a number of controls to prevent them from tangling and knotting. Any failing on the part of these enzymes, which includes re-joining the strands, would be disastrous for the molecule. Even so, the hydrogen bond attractions between the now unpaired bases tends to make the single strands fold back, hydrogen bond to themselves and form unworkable conformations. This is prevented by the action of special single strand binding proteins which bind to the strands and effectively stiffen them in the straight, workable form.

Helicase, starting from the point of origin, moves along the molecule in two opposite directions, unwinding and splitting the strands as it goes. The two separate strands left between the helicases separate and form two replication forks. Each leg of each fork is complimentary, in terms of bases, to the other leg and each leg constitutes a complimentary template against which the new strand will be replicated. The entire replication process works identically and simultaneously on both forks. However, the mode of replication is different for each leg. Fig. 15 represents one of these forks already a considerable way into the process. At the stage we have just considered, the molecule is being unwound, split apart, relieved of stresses, held in straight alignment and the first unpaired parent bases are being exposed behind helicase. Because the parent bases were complimentary and antiparallel to one another, one leg is being exposed in the 3' to 5' direction and the other in the 5' to 3' direction. It is now time for the enzyme polymerase delta to procure the first new complimentary nucleoside triphosphate and catalyse the reaction which attaches it as the first new complimentary nucleotide. However, whilst polymerases are highly efficient in connecting one nucleotide to another in a chain, they are unable to start a chain. This is because there is no free OH group for it to plug into. Also, polymerases can only assemble a chain in the 5' to 3' direction and only one leg of the fork is disposed antiparallel to this. This leg will match what is called the Leading Strand and is the right-hand leg in Fig. 15. The first new nucleotide to be added was the one at the inside bottom of the right-hand leg in Fig. 15. The problem of the missing free OH group into which this was plugged, so to speak, was overcome by the action of the enzyme RNA primase. This enzyme brings in what is called an RNA primer. This is a short piece of single strand RNA complete with the first two complimentary bases which hydrogen bond to the first two parent bases. The free OH group on the 3' carbon of the ribose sugar of the RNA primer serves as the socket into which polymerase delta can then plug the next nucleotide. Replication of the leading strand then carries on

94

continuously in the 5' to 3' direction as shown by the long hollow arrow in Fig. 15. The destabilising forces already mentioned, arising from helicase splitting the molecule apart, also have an effect on polymerases. To overcome this, a specialised set of clamping proteins form a sliding, ring clamp around polymerase delta and the replicating strand so holding them in stable, close, accurate contact. The left-hand leg in Fig. 15 poses some problems for polymerase epsilon. It is being exposed by helicase in a 5' to 3' direction and polymerase, as well as being unable to start a strand, cannot polymerise in the antiparallel 3' to 5' direction. So, the process is delayed on the left-hand leg which, by consequence, is called the Lagging Strand. RNA primase and polymerase epsilon both wait until a short length, two hundred or so bases, of the parent strand have been exposed. RNA primase then inserts an RNA primer and polymerase epsilon, also with a ring clamp, then works backwards in the 5' to 3' direction adding the new nucleotides as shown by the short hollow arrows in Fig. 15. This waiting and forming process is repeated resulting in the left-hand leg being replicated in short lengths called Okazaki Fragments, named after the Japanese man and wife team who elucidated this part of the process. When a specified region of the molecule has been replicated, the enzyme polymerase 2 locates and removes the RNA primers and replaces them with DNA nucleotides. The enzyme DNA ligase then joins the fragments into a continuous chain. As the bubbles on the multiple sites of replication meet, continuous replication has been achieved and the process stops. However, replication cannot reach the very end of the chromosome and if this were ignored, replication would not be complete. To overcome this, the chromosome is lengthened by a repeated sequence of DNA called telomeres. This allows replication to cover the meaningful length of the chromosome.

The speed of replication is very fast, at about three thousand nucleotides per minute, and there are a number of factors that can cause errors to be introduced. The replication process therefore includes proofreading and editing facilities. The first line of defence against errors acts during replication. Whilst polymerase is adding a new nucleotide, it is also checking the previous nucleotide added. In the event that this base pair has been miss-matched, polymerase has the capacity to move back one nucleotide, remove the miss-matched base, digest it and replace it with the proper base. This process achieves accuracy in the order of one faulty base pair in ten million. However, this is still a considerable error incidence and there are other factors which can cause errors. Thymine is a methylated form of uracil and can be subject to spontaneous hydrolysis which changes it back to uracil. A nucleoside can also be subject to spontaneous hydrolysis causing it to lose a base component; tautomeric forms can flip from one state to another. Various toxins can oxidise or methylate base components. The intake of benzene from slow combustion of biomass during cigarette smoking can be particularly lethal and the form of cytosine or thymine bases can be altered by the effect of ultraviolet light. All of these effects disrupt base pairing and affect structural distortions in the DNA molecule. There are a host of specialised enzymes attuned to detecting and repairing these faults. They act following replication and reduce the error incidence to less than one in a trillion base pairs.

A host of other complex and highly specifically targeted enzymes now come into play as the cell enters the mitosis phase. They serve to identify the cell's entire compliment of furniture and machinery, to introduce RNAs to replicate it and to re-dispose it equally into two opposite hemispheres of the cell. A groove develops around the outside wall of the cell. It gradually deepens until the two halves separate or divide into two identical cells, the process of cytokinesis.

Considering your body is comprised of some one hundred trillion cells and that, in the early stages of our lives, cells are replicating by the hundreds of millions, and most of us go through most of our lives in reasonable health, it is as true as doesn't matter to say that replication results in two identical cells. However, occasionally a replicative fault or otherwise induced mistake does escape the cell's correction machinery. If such a mistake survives the cell's next round of replication, then it becomes an in-house characteristic and will persist throughout that person's or organism's lifetime. If the mistake is in a gamete cell, (A male sperm or female egg cell) then it will be passed through sexual reproduction and will persist in that organism's progeny. These mistakes comprise both risks and promises and are an essential characteristic of life as a whole. Without them, life would not evolve and we might still be a bug comprised of only a few atoms squirming in a muddy estuary on a planet of completely bare, rain-washed, sun-bleached rock.

I trust the reader now has, at least, a vague understanding of the composition of the information of which our entire being is an interpretation and the way in which it is perpetuated. However meagre your understanding might be, it is doubtless you must see it as a wonder to behold; even whilst such beholding is constrained commensurate with one's power of imagination. We shall now take a brief overview of the interpretation which, in conjunction with the DNA molecule, is engineered by other nucleic acids, ribonucleic acids, or RNAs.

As we have seen, the DNA molecule is a tremendously dynamic, molecular machine with its own brand of immortality. However, it is locked within the nucleus of the cell, (In bacteria, it is still confined but in a less clearly defined area of the cell known as the nucleolus) as an information store. On the other hand, RNA, which is structurally similar, except that it is single stranded and has thymine replaced with uracil, deoxyribose sugar replaced by ribose, is multiple in function, short lived, much shorter in length, more error prone and motile (Able to move around) within the cell, is highly interactive with other cell components. Within the nucleus, it is copied from DNA. At this stage in the cell cycle, the nucleic membrane becomes permeable to RNAs. This allows the RNA to move away from the nucleus and into the cytoplasm of the cell where it goes to work.

Although I feel reasonably satisfied that I have managed to convey to the lay-reader at least a condensed understanding of the DNA molecule in terms of what it is and what it does, it is with some trepidation that I embark upon attempting to explain Gene Expression. I doubt if there are many or indeed any authors, including those qualified, who would attempt such a thing. However, I have a unique advantage which I trust will turn the trick. Being a retired engineer, I too am a lay-person in this respect. Any modern fourteen-year-old child understands Pythagoras' theorem, the manipulations of basic algebra and how the lever principle works. Yet it's not so long ago since such understanding resided only within the realm of science. It would be sad indeed if the scientific academia were to follow Pythagoras's idea that science should remain a completely institutionalised realm of thought with no sight of the natural beauty it reveals to say nothing of our sense of human achievement in its revelations.

As we have just considered, the coded information in the physical format of the DNA molecule, represents, with the exception of individual environmental effects, our entire being. It is most intriguing and fascinating to consider that, scattered along some two metres of DNA, is information which is interpreted into your notion to have an ice cream or a beer with your friends at the pub, into your capacity to interpret the world around you through the wondrous phenomenon of vision, to be enchanted by the scintillating tinklings of Chopin and moved by the synchronies of Beethoven, to touch, to feel, to smell a flower, to be sad or happy, to love, to anticipate, to imagine, to think and to wonder; to live with enthusiasm.

Professor Richard Dawkins, our most eminent evolutionary scientist, in his "Selfish Gene" concludes that the main purpose in our being is to act as a carrier of the immortal gene. This derives from the evolutionist's premise that life is incidental rather than progressive. It is interesting and somewhat romantic but, to my mind, is a rather diminutive view of the immense majesty of life. If there is purpose in life, surely it must be an aspiration on the part

98

of nature as a whole towards cognisance of its own being. That of which we are the culminate evolved state, not just as a carrier of the gene, which purpose any microbe accomplishes adequately, but in our capacity to know we carry it. I greatly admire Richard Dawkins' work. I owe him much for many exciting hours of enthralling contemplation. However, I cannot help but feel that the evolutionist's abject fear of admitting any approach to anything that might be construed as metaphysical may represent a barrier to their conceiving of aspects of nature which are explosively more subtle than they give it credit for.

On the other side of the Atlantic we have Professor Michael Behe, of similar and equal qualification, and others who believe the complete opposite. They believe that God did indeed create the universe and man. In defiance of Darwinian evolution Michael Behe has a classical argument called Irreducibility. He uses the domestic mouse trap to illustrate the idea. If you take a mouse trap and remove any single part of it, it will no longer work as a mouse trap. So, a mouse trap did not evolve in the way our modern computers have. It arrived through an intellectual preconception of its entire structural configuration and function. He goes on then to describe a particular flagellum which is similarly irreducible and therefore must have been arrived at by will of God rather than having evolved gradually through a series of entities with lesser and evolutionary improving functionality. A flagellum is a bacterial kind of nano-machine which is motile (Or in other words, has a degree of mobility derived from a whipping, lashing, rotating or wiggling tail). Some of them have mechanistic configurations that are utterly fascinating. The one referred to by Michael Behe contains an electric motor which is, component for component, exactly the same as a man-made electric motor and works on a similar principle. It is however, much more sophisticated than the man- made motor. Whilst revolving at some four thousand revolutions a minute, it can affect a dynamic stop in just one quarter of one revolution and immediately go into reverse; something no man-made motor can do.

The irreducibility of this particular flagellum would seem to offer a powerful argument in favour of omnipotent design over evolution. However, we shall see later that, if my unqualified idea regarding the purpose of so called "Junk information" on the DNA molecule is correct, then such a complete entity could be arrived at by evolutionary means through the interpolation of newly accrued fragmentary information with existing information related to other existing flagella. An example of a flagellum we are all aware of is the human sperm.

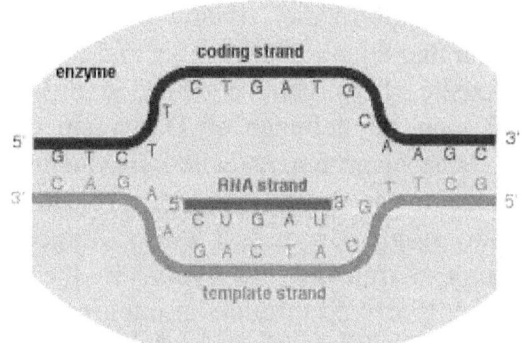

Courtesy Chemguide.Co.uk
Fig.17 Transcription.

Fig.17 shows the DNA molecule in course of being transcribed. You will note that what were called the leading strand and the lagging strand in the DNA synthesis phase are now called the template strand and the coding strand respectively during transcription. The grey template strand in Fig. 17 is equivalent to the outer right-hand leg or leading strand in Fig. 15.

Transcription is the name given to the process of forming a single stranded RNA molecule which is complimentary to a section of the template strand of the DNA molecule. Since the template strand of DNA is complimentary to the coding strand and the new RNA strand will be complimentary to the template strand, the new

100

RNA molecule will be exactly the same as the coding strand which is the outer left-hand leg in Fig.15.

Whilst we shall be considering gene expression in the higher evolved eukaryotes, the process is more fully understood at the prokaryotic level i.e. in bacteria. Therefore, some parts of the processes being mentioned may relate more to bacteria. Again however, as in replication, we shall not differentiate between the two except to emphasise one major difference mentioned earlier. Whereas eukaryotic cells have a clearly defined, membrane bound nucleus, bacterial cells have a less defined unbound area called the nucleolus which has a kind of evanescent mergence into the cytoplasm. In conjunction with this simpler state, the newly formed bacterial RNA is a final product. It does not undergo any post transcriptional modification as it does in the higher evolved eukaryotes. As a result, the new bacterial RNA is ready for immediate translation and indeed, is translated as it is formed. Transcription and translation are coincident processes in bacteria. In eukaryotes, the newly transcribed RNAs are called primary or pre-RNAs and are modified to mature RNAs before leaving the nucleus and entering the cytoplasm.

DNA MOLECULE ONE GENE

5' untranslated region   Coding sequence   3' untranslated region
ACTCCTGAGGAGAAGGTGCATCTGACTCCTGAGGAGAAGGTGCATCTGACTCCTGAG
TGAGGACTCCTCTTCCACGTAGACTGAGGACTCCTCTTCCACGTAGACTGAGGACTC
3'   5'

Transcription------------5' CAUCUGACUCCUGAGGAG 3'   Pre-Messenger RNA
Remove Introns------------ CAUCUG   GAGGAG
Splice Exons---------------   CAUCUGGAGGAG
5' end Cap and Poly A tail | | | | | CAUCUGGAGGAGaaaaaaaaaaaa *Mature*
Messenger RNA
Exit nucleus ----------------- Enter ribosome
Translation-----------------   CAUCUGGAGGAG
Protein----------------------   His   Leu Glu Glu .....Exit ribosome

Fig. 18 Main stages in gene expression.

Fig. 18 shows the main stages in gene expression. The uppermost line of letters is the Coding strand of the DNA molecule and is equivalent to the outer left-hand leg in Fig. 15. The next line is the Template strand and is equivalent to the outer right-hand leg in Fig. 15. The next line down is the transcribed pre-messenger RNA (pre-mRNA). You will note that some parts of these lines are underlined. This is to denote what are called Introns and Exons. The information on the DNA molecule is not continuous. It is interrupted by non-coding sections called Introns. The average human gene has about eight to nine such interruptions with the coding sections called Exons in between. The fourth line down in Fig. 18 shows the introns being cut out and removed. The fifth line shows the exons being spliced together after removal of the introns resulting in a continuous sensible strand of pre-mRNA. The Exons needn't necessarily be spliced in the same order. They can be interpolated and then spliced. This is called Alternative Splicing and widens the range of proteins being coded for by a single gene. Introns and Exons are not regular as shown in Fig.18. This regularity is merely due to a lack of page space. Introns are usually considerably longer than Exons. They can extend from hundreds to just short of half a million base pairs long and constitute the greater part of the DNA molecule. The sixth line down is the final stage before the RNA exit's the nucleus and enters the cytoplasm. The cytoplasm is inhabited by a number of enzymes called ribonucleases specially constructed for the purpose of degrading or destroying RNAs. These enzymes have a double purpose. One is to destroy and get rid of normal RNAs after they have done their job. The other is to destroy foreign, invading RNAs from viruses. Therefore, before normal RNA leaves the nucleus and enters the cytoplasm, it must be protected against these destructive enzymes for long enough for it to do its job. So, before leaving the nucleus, the RNA is fitted out with a special 5' end cap to protect it from enzymes which attack the five prime ends of RNAs. It also has a special PolyA tail attached to the 3' end to protect it from enzymes which attack RNAs from that end. The sixth line down therefore, is now a mature messenger RNA ready to exit the nucleus, enter the cytoplasm and find a ribosome where it will be translated into Protein. The 5' prime and the 3' prime Untranslated regions on the DNA molecule, called the 5' UTR and the 3' UTR, are also copied onto the newly forming RNA. These regions carry sequences which affect a number of regulatory controls throughout the gene expression process. These controls have effects on cell differentiation i.e. will it be a liver cell or a brain cell or a heart cell etc. They also affect cell morphogenesis like, for example, the difference in form between a butterfly and the caterpillar from which it emerges or a frog from its tadpole. They also affect the phenotype emerging from the genotype. For example, a black kitten and a black and white kitten of the same parentage are of the same genotype but they differ in phenotype. This is because of slight differences in the way their genes were expressed. Also, if one of the kittens grows up as a mouser on a farm and the other as a house pet, they will develop different behaviourisms due to environmental effects which will further widen their difference in phenotype. This is a major factor in the development of different species over evolutionary time spans.

The gene transcribed in Fig. 18 resulted in the polymerisation of messenger RNA which, in turn, would go on to be translated into a protein. However, it might just as well have been a non-protein coding gene which would have produced any one of a number of different RNAs including ribosomal RNA (rRNA) or transfer RNA (tRNA) or any one of a considerable number of other RNAs. The 5' and the 3' untranslated regions of these sequences are also copied as part of the RNAs and affect regulatory controls on both the rate and the way in which the RNAs interact within the gene expression process. We shall encounter some of these in the gene expression mechanism which we shall now briefly consider.

The gene expression mechanism, even as an overview, may seem rather complex. However, even if this does not provide a clear picture of the process, please do not misjudge your acumen or mine. Unless you happen to be a genius, subjects of such complexity require detailed explanation and constant involvement in order to generate and maintain clear understanding. You will, at least, gain some appreciation of the intricate and ingeniously interacting complexity of the process.

The reader should bear in mind that our considerations here comprise only a briefest possible overview of what is an immensely complex process. Also, the activities mentioned are chemical reactions triggered, for the most part, by highly specialised and precisely targeted enzymes. Furthermore, material conglomerates are being controlled by specialised functioning proteins and the overall activity is occurring within a space some twenty-five to forty thousand times smaller than a small fractional millimetre length of a single human hair.

The object of gene expression is to convert information on the DNA molecule into a polypeptide chain of amino acids and to affect their subsequent folding into a functional protein. In course of doing this, other chains i.e. ribonucleic acids or RNAs will be formed. These will be formed by the action of already existing RNAs within an interdependent circular process. This answers one of your vital questions before you ask it, "Where do the RNAs that make the new RNAs come from?"

Some of the RNAs being mentioned are named in accordance with an S value like for example, 4S and 18S. These names are derived from the levels at which these RNAs sediment during high speed centrifugation. For our purposes, we can consider the S value like a shoe size, the higher the S value the bigger the molecule. These molecules are categorised mainly by their size
.

Transcription

Following Replication, gene expression occurs in two further, main stages called Transcription and Translation which, in themselves, comprise a number of interim stages. It begins with transcription and transcription begins in a manner similar to that of replication. It is driven by three main RNA polymerases POL 1, POL 2 and POL 3. You will remember, from replication, that polymerases join molecules into polypeptide chains. These polymerases have some shared functions and are aided and abetted by a large number of other functional RNAs and ancillaries such as Binding Proteins called Transcription Factors together with already prepared sites on the DNA molecule. These sites facilitate targeting and are highly receptive to specific enzyme and binding protein activity. They are called Promoters.

Transcription can be considered as occurring in five stages which are: pre-initiation, initiation, promoter clearance, polymerisation and termination

.

Pre-initiation:

A core promoter is needed to start the process and, in eukaryotes, this lies in a region of the DNA molecule 25 to 30 base pairs upstream from the Transcription Start Site. It is comprised of a short sequence of thymines and adenines as TATAAA and is called a tata box. The tata box is a core promoter and acts as a binding site for a transcription factor called Tata binding protein. The Tata binding protein is a subunit, or part, of another transcription factor called transcription factor II D. When transcription factor II D binds to the Tata box through the action of Tata binding protein, five more transcription factors combine around the Tata box in a series of stages to form a pre-initiation complex.

Initiation

:

RNA polymerase Pol II, which does not directly recognise the core promoter, can, by the action of the transcription factors now in place, bind to it. This completes the transcription initiation complex which appears as a bubble on the DNA molecule. During transcription, several or many initiation complexes can be operative on a gene at the same time producing multiple transcripts of a single gene.

Each of the RNA polymerases, Pol 1, Pol 2 and Pol 3 is a complex, multi subunit conglomerate which, in conjunction with the transcription initiation complex, carries a complete compliment of machinery similar to that contained within the replication bubble. However, since, in transcription, only one gene at a time is being transcribed, the bubble, rather than expanding in both directions, moves in a single direction along the DNA molecule. The three RNA polymerases have distinct, but to some extent overlapping, rolls in the transcription process.

RNA Pol 1 is comprised of twelve subunits in humans. It is responsible for transcribing all of the ribosomal RNAs except one, the 5S. It transcribes one large transcript, encoding a ribosomal gene over and over again. This gene encodes the 5.8S, the 18S and the 28S ribosomal RNAs (rRNAs) The transcripts are separated by another RNA called snoRNA and will go on to form parts of a ribosome.

RNA Pol 2 is also a twelve-subunit complex and catalyses the transcription of DNA to synthesise pre-messenger RNA (pre mRNA). It also synthesises snRNA and microRNA. It requires a wide range of transcription factors (Some fifty in all), to assist in binding to its promoters to begin transcription.

RNA Pol 3 transcribes the synthesis of transfer RNA (tRNA). It also transcribes the 5S component of ribosomal rRNA and some other small RNAs. The genes transcribed by Pol 3 are called housekeeping genes because their expression is required in all cell types. The regulation of Pol 3 transcription is therefore tied to regulation of cell growth and cell cycle.

The actions of the three RNA polymerases on the template strand of the DNA molecule are fairly similar so we shall confine ourselves to the action of Pol 2 which polymerises pre-messenger RNA.

Polymerisation.

Once bound to its promoter, the Pol 2 complex which contains helicase activity starts to unwind the DNA helix. The topoisomerases come into action, as in replication, and relieve the stresses caused by unwinding. Pol 2 is also associated with what is called a Sigma Factor. A Sigma Factor is a site recognition, organising and binding protein. Pol 2 now procures the first appropriate RNA nucleotide and attaches it to its complimentary counterpart on the template strand of the DNA molecule just as DNA polymerase did with the RNA primer during replication. Pol 2 is reading the DNA template strand in its 3' to 5' direction and will assemble the new RNA strand anti-parallel in the 5' to 3' direction just like the socket and plug arrangement we considered in replication. Having placed the first RNA nucleotide, which hydrogen bonds to its complimentary counterpart on the template strand, Pol 2 now clears the promoter in order to move to the next position. This causes a tendency for the new transcript to be released causing shortened or incomplete transcripts. This is called Abortive Initiation and the condition is rectified by action of the sigma factor. The sigma factor stabilises the process over a distance of thirty-five base pairs and the stabilisation moves along with Pol 2 as it progresses. When the transcript is about twenty-four nucleotides long, the process is stable and the sigma factor is released a little later. Polymerisation of the new mRNA strand can now continue uninterrupted in the direction of the long, hollow arrow in Fig. 15. Pol 2 procures the nucleotides as ribonucleic adenosine triphosphate nucleosides. They are hydrogen bonded to their complimentary counterparts on the DNA template strand and the sugars bonded by the reduction of the adenosine triphosphate in the 5' to 3' direction just as in replication.

Transcription also includes a proof-reading and correction mechanism but it is not nearly as sophisticated as that in replication. When Pol 2 moves from one nucleotide to the next, it checks that the previous nucleotide just added is correctly base paired. If it is not, it removes the faulty nucleotide together with the one immediately previous and then replaces both of them.

Termination.

When the transcript has been completed, the new, single stranded RNA molecule, which is hydrogen bonded to the DNA molecule by complimentary base pairing, has to be released from the DNA. This is called Termination and this process is not yet fully understood in eukaryotes. However, it has been elucidated in bacteria which use two different methods. One is called Rho-dependent and the other Rho-independent transcription termination. In Rho-independent transcription termination, RNA transcription stops when the newly synthesised RNA molecule forms a guanine-cytosine rich hairpin loop followed by a run of uracils. When the hairpin forms, the mechanical stresses break the bonds between the RNA and the DNA. This pulls the uracils out of the active site of Pol 2 and terminates transcription so allowing the new RNA molecule to move away from the DNA.

The Rho factor is an open ring enzyme capable of messenger RNA recognition and has RNA/DNA helicase activity. In Rho-dependent transcription termination, the Rho factor destabilises the hydrogen bond interactions between the DNA template strand and the newly formed messenger RNA so releasing the messenger RNA from the polymerising complex.

The next and final stage of gene expression is Translation. However, before going on to translation, it is worth pondering for a moment on a particular relationship between replication and transcription which you have undoubtedly already noticed. You will remember that, during replication, the error detection and repair mechanisms are numerous and highly sophisticated to the extent of being almost infallible. On the other hand, the equivalent mechanism in the transcription process is more or less singular and crude to say the least. These mechanisms do not arise as an automatic consequence of the already ongoing chemistry. They have been introduced into the processes in terms of their specific functions i.e. error detection and repair and operate at a cost in energy usage. You will also remember that, during replication, any error is likely to be serious in the respect that it may persist throughout the lifetime of the organism affected through further subsequent replication. Worse still, if the error happens to be in a gamete cell, it will persist through that organism's progeny. On the other hand, an error in transcription can be self-abortive because of the way in which transcription works, but even if it persists, it will only affect a cell which has a lifetime of only months, weeks or even days. So, life seems to have made a judgement here in terms of risk / cost ratio. This is exactly the same, and on the same basis, as judgements that have to be made by designing engineers and architects every day. In these terms, replication might be likened to an engineer designing a manned orbital space vessel or passenger jet engine and transcription to one designing a DIY cordless drilling machine or bicycle pump. It is another instance where life seems to lean heavily in the direction of teleological rather than teleonomic concept.

Translation.

Translation is the process whereby information on the DNA molecule, having been transcribed into messenger RNA, is then converted into a polypeptide chain of amino acids. These in turn roll and fold into functional proteins which, together with water, fats, starches and a minimal amount of a few other substances comprise the living stuff of which you are made.

The active machinery in translation is called a ribosome. It works in conjunction with amino acid carrying transfer RNAs. Every amino acid has one or more associated transfer RNAs which work in conjunction with a special carrier and elongation enzyme called the eEF-1 factor. Avoiding the technical chemistry, if we consider each transfer RNA and its associated amino acid as a pair, this pairing is affected by action of the special enzyme which connects them together. This enzyme is called Aminoacyl-tRNA

Synthetase. There is an individual aminoacyl-tRNA synthetase associated, by means of special recognition factors, with every individual transfer RNA and its associated amino acid. It is by this means that every transfer RNA is connected to its proper amino acid, coinciding with its anticodon.

There are a considerable number of ribosomes floating around in the cytosol and also a number attached to and causing the roughness on the rough endoplasmic reticulum. A ribosome is a small but hugely complex un-membrane bound organelle. It is comprised of two subunits, a larger and a smaller together with a considerable number of ribozymes, (A ribozyme is an RNA which has some of the properties of a protein enzyme). Ribosomes floating freely in the cytosol are responsible mainly for the synthesis of in-cell proteins. Those attached to the endoplasmic reticulum make proteins mainly for export out of the cell assisted by the Golgi apparatus. The Golgi apparatus modifies, packages, labels and distributes proteins. It is sometimes referred to as the cell's post office. (All of these other cell components being mentioned are produced by the same sequence of replication, transcription and translation we are considering here) A ribosome, during the process of translation, can respond to signalling mechanisms which direct it to attach to the endoplasmic reticulum. Following translation of the protein coded for, it can then return to the cytosol and translate other in-cell proteins.

During translation in eukaryotes, the messenger RNA, complete with its coded instructions for making a protein, interacts with the nucleic membrane pore complex in such a way as to pass through it out of the nucleus and into the cytosol (The active semi liquid component of the cell). From there, it finds, enters and becomes embedded within the smaller subunit of a ribosome. Remember that the messenger RNA is comprised of a 5' untranslated region, a protein coding region or Open Reading Frame and a 3' untranslated region. At this stage, the ribosome is attempting to read the 5' untranslated region. (The lay reader, by this time, will have become used to DNA and RNA being read in

110

the 3' to 5' direction and polymerised in the 5' to 3' direction because of their plug and socket arrangement. But the smaller subunit of the ribosome is not a polymerising mechanism so it reads the messenger RNA in the 5' to 3' direction starting with the 5' untranslated region). This region does not code for amino acids so the ribosome cannot read it. So, the ribosome moves along the RNA molecule until it finds a Start codon. The start codon is AUG and AUG has two functions. It signals the start of the open reading frame and it also codes for the amino acid methionine. Accordingly, every chain of amino acids forming a protein begins with the amino acid methionine. In proteins where methionine is not to be included in this position, it is subsequently removed. If AUG occurs somewhere else in the middle of a reading frame it does not signal start. It merely codes for the amino acid methionine. The larger subunit now connects to the smaller and the smaller subunit proceeds to read the messenger three-bit words, called codons one at a time. Each codon codes for a particular amino acid. It is utterly essential that the ribosome finds the exact abridgement of the AUG codon. If it is only one bit out in either direction and then proceeds to move three bits at a time, the whole reading frame will have been altered. It will result in the production of a different and dysfunctional protein. The movement and positioning of the smaller subunit along and on the messenger RNA is affected by another enzyme called the eEF-2 factor. The eEF-1 protein, working in conjunction with the smaller subunit, selects and delivers a transfer RNA complete with the anticodon to the codon being read and its associated amino acid into the now combined smaller and larger subunits of the ribosome. The first amino acid to arrive is always methionine. The larger subunit stabilises the transfer RNA and ensures that its amino acid is properly oriented so that its amino end can react with the carboxyl end of the next incoming amino acid to form a peptide bond. This is a normal condensation reaction as described earlier. It is activated by the larger subunit so that the hydrogen on the trailing amino end of the sitting amino acid joins the hydrogen / oxygen of the carboxyl group on the leading end of the incoming amino acid and forms a molecule of water which drops out. In the same reaction, the nitrogen of the amino end of the sitting amino acid is covalently bonded with the carbon of the carboxyl end of the incoming amino acid thus bonding the two amino acids together.

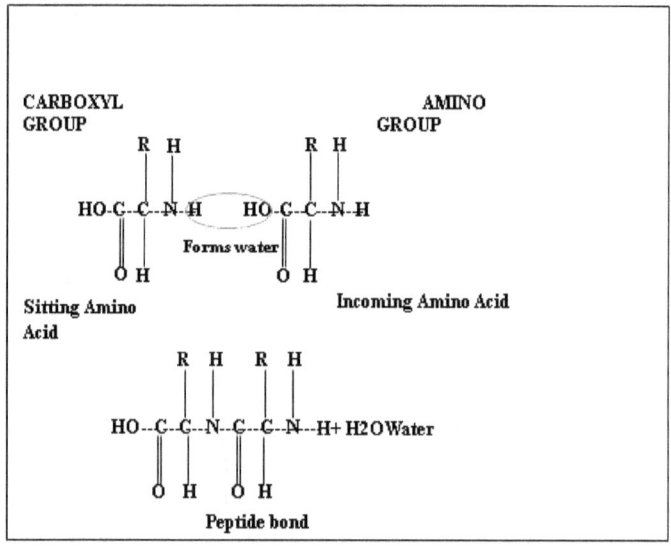

Fig. 19 Joining of amino acids within larger subunit of ribosome.

Steps in Translation

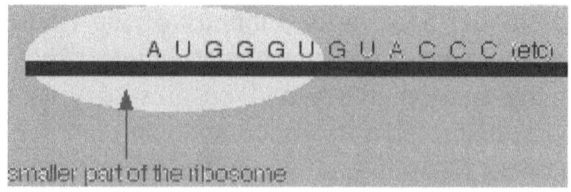

Courtesy Chemguide.co.uk

Fig. 20 A Messenger RNA enters small subunit of ribosome

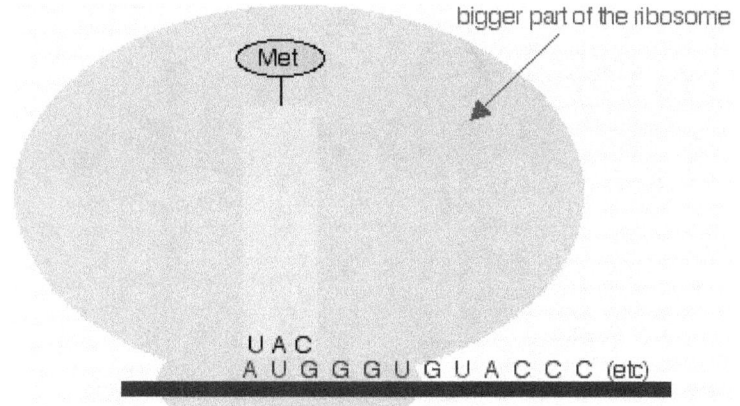

Courtesy Chemguide.co.uk

Fig. 20 B Larger subunit connects to smaller. First transfer RNA complete with anticodon and amino acid arrives.

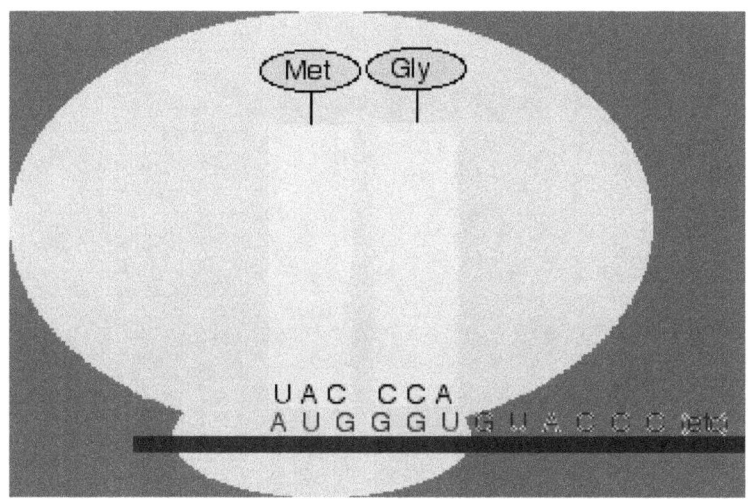

Courtesy Chemguide.co.uk

Fig. 20 C Second transfer RNA arrives. Amino acids bonded together.

113

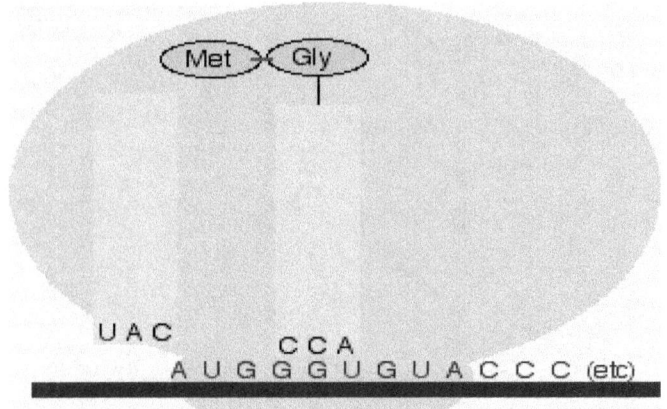

Courtesy Chemguide.co.uk
Fig. 20 D First transfer RNA, now unloaded, released back to cytosol.

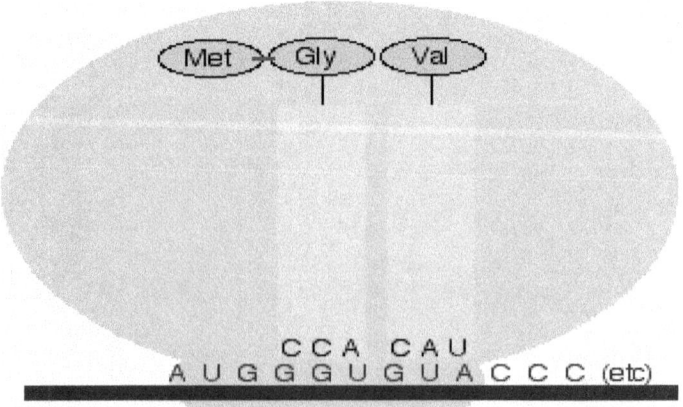

Courtesy Chemguide.co.uk
Fig. 20 E Third transfer RNA arrives. Amino acids bonded together.

114

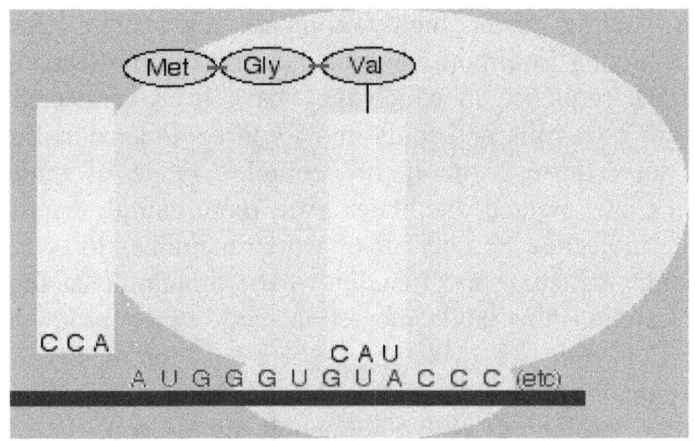

Courtesy Chemguide.co.uk

Fig. 20 F Second transfer RNA released back to cytosol.

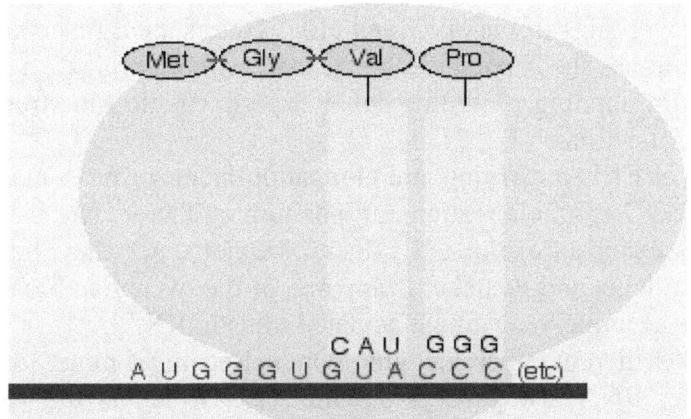

Courtesy Chemguide.co.uk

Fig. 20 G Fourth transfer RNA arrives. Amino acids bonded together.

This process continues until the ribosome encounters any one of three stop codons. These are UAA, UAG and UGA. These do not code for amino acids so the process stops and the linear polypeptide chain of amino acids produced is released from the ribosome. Acting under highly complex and subtle chemical influences arising from both the molecular make-up of each amino acid and the sequence in which they have been connected, this linear chain now rolls and folds into its three-dimensional shape. This, in turn, through the highly complex range of molecular proximities thus created, facilitates even more complex and more subtle intermolecular activity all of which combines to determine the particular structure and function of the protein. That function includes relationships and interaction with other proteins and substances in an ever-widening sphere of complexity which ultimately combines into the sum of all that you are including how you relate to and interact with other people around you.

If you have been following this process in your mind and are aware that the eEF1 factor was not inside the ribosome at the start of translation, you may be wondering how it knew the sequence in which to deliver the transfer RNAs.

The workings of the eEF1 and eEF2 factors are a fine example of how life has the capacity to be "Clever without brains"; a result of constant ongoing multidirectional experimentation in structural form and function.

The eEF1 is a carrying and elongation factor or nano-machine and the eEF2 is an elongation and positioning factor. Just like you going to the theatre, the eEF1 factor cannot enter the ribosome without a ticket and its ticket is any one of the twenty amino acids. (Or more accurately, an amino acylated transfer RNA).

You will remember that the "Open Reading Frame" on the messenger RNA always begins with AUG which is the "Start" signal and is also the codon for methionine. Because of this, when the eEF1 is first activated, it does not need to read the first or Start codon. Part of its initial activation signal programmes it to collect and deliver the transfer RNA carrying methionine; its ticket to enter the ribosome. When it enters the smaller sub-unit of the ribosome, carrying the methionine, the eEf2 factor aligns the anti-codon for the methionine with its codon on the messenger. This automatically aligns the abridgement of the eEF1 factor with the next codon in sequence. Whilst the methionine is being unloaded, the eEF1 is reading and being re-programmed by this next codon. It then moves back into the cytosol and collects the next amino acid carrying transfer RNA in sequence. This process continues until the eEF1 encounters a stop signal on the messenger RNA. The smaller subunit of the ribosome is therefore the receiving, reading, alignment and sequencing instrument for the assemblage of amino acids into proteins which is carried out by the larger subunit.

The amino acids of which the protein is comprised emerge from the larger subunit of the ribosome in a straight, linear polypeptide chain. This then proceeds to roll and fold into what seems a tangled, balled conglomerate of protein substance much like a tangled, knotted ball of string. However, the way in which the amino acids are rolled and folded is not a random melee. The particular way in which it is rolled and folded is as important to the form and function of the protein as the kind and sequence of the amino acids of which it is comprised. If this seemingly mixed up rolling and folding is interfered with in any way, it results in a dysfunctional protein. Clearly then, the passage of messenger RNA through the ribosome, as well as determining the kind and sequence of amino acids forming the polypeptide chain, also imparts some kind of subtle, genetic information over the length of the chain which determines the particular way in which it rolls and folds. It would seem that this, in turn, determines a particular pattern of molecular proximities which then interact in such a way as to determine the substance, character, form and function of the protein. This is another utterly fascinating aspect of proteins which is presently a subject of intense research in attempting to understand it.

My own personal, unqualified feelings on this aspect of protein structure are, that certain degrees and angles of proximity between certain types of molecules invokes a kind of reaction of which, as yet, science knows absolutely nothing about. It is a reactionary force which science cannot yet detect or measure. It may even be a kind of negative energy which is oriented towards order rather than disorder. I believe that this force expressed itself in the assembly of the first nucleic acids and carries on through the entire life process to its culminate expression in humanity. In other words, it is the essence of life. Mumbo jumbo you might call this, but try to imagine the workings of your own imagination.

The following table is coloured to illustrate the synonymous groupings of amino acid codons.

second base in codon

		U	C	A	G	
		UUU Phe	UCU Ser	UAU Tyr	UGU Cys	U
	U	UUC Phe	UCC Ser	UAC Tyr	UGC Cys	C
		UUA Leu	UCA Ser	UAA stop	UGA stop	A
		UUG Leu	UCG Ser	UAG stop	UGG Trp	G
first base in codon	C	CUU Leu	CCU Pro	CAU His	CGU Arg	U
		CUC Leu	CCC Pro	CAC His	CGC Arg	C
		CUA Leu	CCA Pro	CAA Gln	CGA Arg	A
		CUG Leu	CCG Pro	CAG Gln	CGG Arg	G
	A	AUU Ile	ACU Thr	AAU Asn	AGU Ser	U
		AUC Ile	ACC Thr	AAC Asn	AGC Ser	C
		AUA Ile	ACA Thr	AAA Lys	AGA Arg	A
		AUG Met	ACG Thr	AAG Lys	AGG Arg	G
	G	GUU Val	GCU Ala	GAU Asp	GGU Gly	U
		GUC Val	GCC Ala	GAC Asp	GGC Gly	C
		GUA Val	GCA Ala	GAA Glu	GGA Gly	A
		GUG Val	GCG Ala	GAG Glu	GGG Gly	G

(third base in codon)

Courtesy Chemguide.co.uk.
Table 1. Synonymous groupings of RNA codons.

You can see, from table 1, that all of the amino acids except two are represented by more than one codon. The exceptions are methionine which is also the start codon and tryptophan. However, in most organisms, a particular single codon for a particular amino acid seems to be favoured. This is the result of a purely arithmetic relationship. With twenty amino acids coded by four nucleotides, a three-bit code is necessary. A two-bit code could only code for sixteen amino acids. A three-bit code can accommodate sixty-four. Also, there are considerably more transfer RNAs than amino acids so an amino acid can have a number of transfer RNAs associated with it as codon carriers.

Chapter 5 - Some Other Amazing Cellular Activities

All of the cellular activity we have thus far considered constitutes only a small part of the overall cell metabolism. This goes on within its own special environment which is bounded and individualised by the cell membrane. The cell membrane is not just a packaging containing the cell contents. It is one of the more important functional organelles forming part of the cell. As well as communicating with and interacting with other cells to form tissues and other structures, it controls and sustains the in-cell environment, the constitution of the cytosol and cytoplasm and is selectively bi-permeable, admitting ingress of particular nutrients from the outside environment and ejecting the waste products of the cell metabolism. The means by which it does this are intriguingly ingenious. The membranes enclosing and compartmentalising organelles within the cell, such as the nucleus, Golgi apparatus, mitochondria and reticulums are similar in both constitution and function to the main cell membrane but are respectively characterised by some differences in structure and functional proteins. Cell types are also differentiated in a similar manner.

In the cell membrane, life utilises some non-energy based, naturally occurring phenomenon such as immiscibility of substances, concentration gradients and osmosis. This accounts for about half of the activity going on in the membrane. The other half is comprised of some ingenious energy dependent mechanisms powered by the same ATP to ADP reaction powering some of the processes discussed earlier. Immiscibility merely means the refusal of some substances to mix together, like water and oil for example. A concentration gradient is exactly what it says it is, a variation in concentration. If you use a foot pump to inflate your car tyre or a hand pump to inflate your bicycle tyre you will be immediately aware that you need to supply a significant amount of energy. When your tyre is at the required pressure, it now contains much more air than an equivalent volume of space at normal atmospheric pressure. In other words, there is a much higher concentration of air inside the tyre than outside so a concentration gradient exists across the wall of the tyre. You had to use energy to push the air in the tyre up the concentration gradient. If you now remove the valve from the tyre, there will be a swoosh of air hissing out of the tyre as the air falls down its concentration gradient towards equilibrium with the outside air. No energy is required here; it just falls down the concentration gradient like a ball rolling down a hill.

Osmosis is a little more difficult to explain. This is because, whilst it is a very common, every day, natural phenomenon, it is one that seems to go against our natural intuitive way of thinking. You will be well aware of the saying, "Water always finds its own level". This is true of course but water's own level is not always level if the water is the solvent in a solution compartmentalised by a membrane tuned to admittance of water molecules and rejection of the solute molecules. Any explanation of how osmosis actually works in terms of kinetic energy, entropy and other calculable aspects is far beyond our considerations here. Even though it is sufficiently understood to be utilisable as an applied science, the fundamentals of its workings are still a subject of some debate in the scientific community. However, it is the means by which life maintains the shape and rigidity of the cell in all life forms and accounts for the degree of firmness in leaves and non-woody plant stems. I shall attempt therefore to describe osmosis by experiment in terms of undesignated units and, hopefully, provide the lay reader with a reasonable understanding of what it is and what it does. You needn't carry out the experiment unless you are sufficiently interested to see the proof and seeing the proof in the experiment I am about to describe might be rather difficult. The experiment requires apparatus comprising two membrane divided U tubes standing exactly vertical, two marked beakers and a very small pot powder measure. Consider a clear glass U tube designated X as in Fig.21 A and which is equally divided by a water tuned membrane i.e. a membrane which will admit passage of water molecules through it and will prevent passage of solute molecules i.e. whatever is dissolved in the water. We shall call the left arm of the U tube "Arm A" and

the right arm "Arm B". Now use a marked beaker to measure out a quantity of sixty units of pure, clean water and pour it into either end of the U tube and let it settle. Water will pass through the membrane until it settles exactly level with an equal amount of thirty units of water on each side of the membrane. We shall call this "Level 1".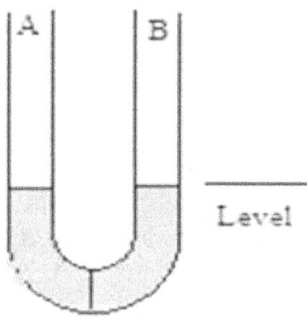

Fig. 21 A U tube X Level 1

[OBJ] Due to temperature dependent kinetic energy, water will flow through the membrane equally in both directions maintaining the liquid at level 1 in both arms of the U tube. In terms of osmosis, this is called an isotonic condition.

Now take the small pot measure and, one at a time, put fifteen units of table salt into each arm of the U tube. Give it a shake to dissolve the salt. You will now find that the water level has lifted slightly to level 2 because you have increased the volume of the

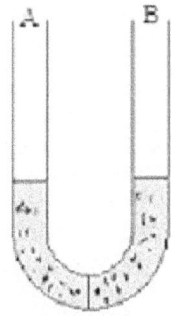

liquid by adding salt. However, the liquid will still settle exactly level. There is an equal concentration of solvent i.e. water and solute i.e. salt on both sides of the membrane. This is still an isotonic condition

Fig. 21 B   U tube X Level 2

Now take another exactly similar U tube designated Y and two marked beakers. Measure out a quantity of thirty units of water in each beaker. Take the pot measure and put

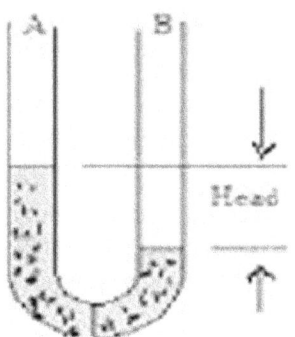

twenty units of salt into one beaker and ten units into the other. Stir and shake to dissolve. Pour the beaker with twenty units of solute into arm A and the one with ten units into arm B then have lunch while it settles. On your return, you will find that the liquid level has risen in arm A to above level 2 and has fallen to below level 1 in arm B. The difference, with the units used

Fig. 21 C U tube Y

in this experiment, will be extremely marginal but the surfaces of the liquid in both arms of the U tube are no longer level.
So why does this happen? It is generally accepted that it is caused by a redistribution of free energy deriving of kinetic energy in the solvent, in this case, the water. This causes the water to appropriate an equal number of water molecules to every molecule of the solute, in this case, the salt. In our U tube Y, there is a higher concentration, twenty units, of solute in arm A than in arm B which

123

has only ten units. Arm A would be said to be hypertonic relative to arm B and arm B hypotonic relative to arm A. Initially therefore, there are many more molecules of water per molecule of salt in arm B than in arm A. If there were no membrane or if the salt molecules could pass through the membrane, the liquid would assume a perfect level as in U tube X. However, there is a membrane and the salt molecules cannot pass through it but the water can. In order to equalise its free energy therefore, water will flow through the membrane from arm B to arm A until there is an equal number of water molecules per unit of salt on both sides of the membrane thus bringing the free energy of the solvent to a state of equilibrium. In terms of concentration ratio between solvent and solute, the water has moved down its concentration gradient. However, in doing this, there is now a higher volume of water in arm A than in arm B. So, the water level in arm B has gone down and the level in arm A has gone up. There is now a "Head" or pressure of water in arm A and the force supporting this head is called the osmotic pressure. It is this which supports the shape and rigidity of cells, leaves and plant stems. Whilst this is only a partial explanation, it is, nevertheless, a reasonable lay person's understanding of what osmosis is. It is utterly crucial to your life and anything interfering with it can have drastic effects. If, for example, (With a few exceptions which have specialised controlling mechanisms) you place a live sea living fish in a tank of fresh water, bearing in mind the fish breathes water, the higher concentration of salt in its cells will induce intake of the fresh water to the extent that it's cells will rupture, killing the fish. If you place a river fish in sea water, the higher concentration of salt outside its cells will cause its cells to exude water. The fish will become limp and flaccid and die. I plead with you not to experiment with this. It is the cruellest possible way for a fish to die.

Osmosis is a naturally occurring phenomenon and is utilised by the cell membrane which is a much more complex membrane than the simple one in our U tubes. As well as osmosis, the cell membrane also utilises passive and ATP - ADP powered protein functions some of which act to control the osmosis effect. The membrane is selectively bi-permeable i.e. allowing selective passage in both directions, to a range of substances in accordance with the requirements of the cell metabolism and productivity. It is another of life's many fascinating wonders of biological engineering and goes on in a highly controlled manner instant to instant in some one hundred trillion cells throughout your body every minute of your life.

The cell membrane, often called the plasma membrane, is an integral feature surrounding and individualising every living cell. As well as controlling the passage of substances in and out of the cell, it is the medium across which the cell's electrical potential is developed and maintained. It also has the capacity, by means of signalling mechanisms and special binding sites, to interact both with other organelles within the cell and with other complete cells outside. You can easily test the effectiveness of this. Give your big toe a single wiggle. The intention to wiggle arises in your brain and your toe wiggles more or less immediately (In three hundredths of a second) and, by return, your brain is just as immediately aware of and has interpreted your perception of the wiggle. (Incidentally, if your big toe develops a habit of wiggling on its own, see your doctor immediately. This is a sign of motor neuron or brain disease.) The cell membrane is comprised essentially of what is called a phospholipid bilayer together with proteins and carbohydrates standing in a relatively sparse supporting and temperature sensitive core of cholesterol. The proteins are embedded in the phospholipids forming what has become known as a fluid mosaic structure in which the proteins can easily float around much like blocks of wood in a pond. Lipids are fats which, as you know, are insoluble in water. A phospholipid is a molecule comprised of a non-polar fatty acid chain at one end and a polar glycerol-phosphate-nitrogen group at the other. Phospholipids tend to self-arrange themselves into two layers (Hence bilayer) usually with the non-polar fatty acid chains facing each other in the middle and the polar phosphate heads facing outwards. This polar head is

125

hydrophilic, which means it can interact with water molecules, whilst the non-polar tails are hydrophobic i.e. repel water.

A typical example of a lipid bilayer is a soap bubble. However, the soap bubble is quite different from the cell membrane. In the soap bubble, the polar heads face inwards and the non-polar tails face outwards. This is because the film of water comprising the bubble is in the middle. Also, the film of the soap bubble is much thicker. If a complete dictionary were used to represent the thickness of the wall of a soap bubble, the thickness of the cell membrane would be represented by the thickness of only one page. The cell contains water and, as a whole, exists within a watery medium so there is water both inside and outside the cell. The phospholipids of the cell membrane therefore, arrange themselves so that the polar heads face inwards to the cell and outwards to the surrounding medium with the non-polar tails facing each other in the middle, the complete opposite of the soap bubble.

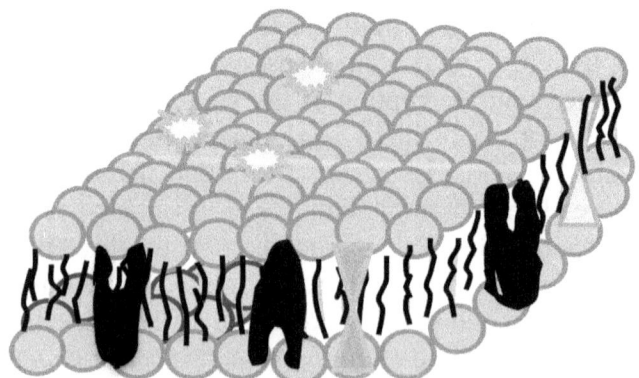

Fig. 22 Small section of cell membrane

We have only considered a relatively small amount of the activities going on within the cell but the overall cell activity requires the intake of water together with a range of substances as raw materials for its manufacturing processes. This, in turn, results in the production of a considerable amount of functional entities as

126

well as waste by-products. In relation to the watery medium in which the cell as a whole exists therefore, there is usually a much higher concentration of solutes inside the cell than outside. Having seen how osmosis works and the fact that it is a naturally ongoing process that cannot be switched on or off, one can immediately see that the cell, whilst requiring the osmosis effect to maintain its internal aquatic environment, must, at the same time, be the subject of constant osmosis attack. Life has solved this crucial problem by utilising the natural phenomenon of immiscibility and evolving some ingeniously subtle protein mechanisms which are embedded in the fluid mosaic structure of the membrane. The polar, hydrophilic phosphate heads of the phospholipids on the outside face of the membrane face outwards into the watery medium surrounding the cell and are compatible with that medium. The polar, hydrophilic phosphate heads of the phospholipids on the inside face of the membrane are in contact with the internal water content of the cell and are just as happy. In between these two hydrophilic layers lies the extending hydrophobic, fatty acid tails forming a region which is incompatible with water. This region constitutes an effective water seal surrounding the cell. However, it is not one hundred percent efficient and does allow a degree of seepage of small molecules including water molecules at different rates dependent on variations of concentration gradient. The seal is sufficiently effective however, to maintain the aquatic pressure of the cell within its functional parameters for considerable periods of time. When these parameters are threatened, a set of ingeniously constructed proteins called aquaporins comes into play. Aquaporins are embedded, transmembrane channel proteins. They can admit passage of water molecules in either direction and at controlled rates across the membrane. Their main attribute however, is that whilst they utilise the osmosis effect to power the water transport (This is called "Passive transport".) they can assume controlled states that are open or closed to water transport. They therefore constitute the cell's defence against osmosis attack. This is an evolved but highly sophisticated technology now copied by industry in a wide range of separation processes.

The water molecule is relatively small and determines the optimal molecular size admissible by the membrane. Consequently, other small molecules such as lipid soluble steroids or oxygen and carbon dioxide can freely diffuse through the membrane. This is called "Lipid diffusion". At the same time, the cell requires admittance of much larger molecules to satisfy its metabolic and manufacturing needs. These needs are met through the action of other ingeniously constructed and specifically functional proteins. Channel Proteins, of which the aquaporins already mentioned are one kind, form a channel through the membrane. They allow charged substances such as ions to diffuse across the membrane. They can be controlled open or closed so the cell can control the entry and exit of ions. Also, there are a considerable number of different "Carrier Proteins". Each different type has a specific binding site for a specific substance. It can change between two states where the binding site is open alternately to opposite sides of the membrane. The substance binds on the side where it is at high concentration and is released on the side where it is at low concentration. These are what are referred to as "Passive transport" systems since they follow natural concentration gradients and do not require the cell to expend energy.

The cell membrane also has the capacity to transport larger substances. This requires the use of cellular energy and is called "Active transport" The means by which it does this are utterly fascinating. One of these means is by "Vesicles" and is fascinating because of its peculiarity. Another is "Pumping" and is fascinating in terms of its molecular, mechanistic technology. Vesicles are capable of transporting very large molecules such as proteins, polysaccharides, nucleotides and even whole cells across the membrane. (A typical example of this is a white blood cell devouring a bacterial cell) A vesicle can develop on either the inside or outside surface of the membrane. It begins by the development of a small dent or hollow on the membrane surface which attracts the particle to be transported. Once the particle is attached, the dent further develops into a pit or mouth and when it is sufficiently deep to contain the particle, the surface of the membrane completely closes so that the particle is wholly engulfed. The vesicle then forms a transmembrane channel through which it forces the acquired particle whilst, at the same time, digesting it. It then opens on the opposite side of the membrane and ejects the digested products into or out of the cell. The action of a vesicle might be compared to a large red apple stuffed tightly into the foot of a child's Christmas stocking. In removing the apple, the child turns the stocking inside out. This is not quite but is something like the way the vesicle works by changing shape like the stocking. A vesicle might be considered as what science fiction writers now call a shape shifter.

The membrane also contains a large number of special proteins which act as pumps to pump specific substances across it and we shall briefly consider one of the more important of these. These are called pumps because they can force substances against their concentration gradients just like you inflating your tyre. The membrane has an electrical potential existing across it. This is another way of saying it has a voltage across it just like the voltage across the two terminals of a battery. Depending on cell type, the membrane's voltage varies between 20 and 200 millivolts i.e. thousandths of a volt which is used to power the cell's signalling mechanisms. This voltage is achieved and maintained through a particular distribution of ions and is always negative inside the cell and positive outside. You will remember from our consideration of chemical compounds that salt dissolved in water produced positive sodium ions and negative chlorine ions. You will appreciate then that a higher concentration of positive ions than negative ions on one side of the membrane and a higher concentration of negative ions than positive ions on the other will produce an electrical potential across it. The pump we are about to consider does not use chlorine however; it uses sodium and potassium ions, both of which have a positive charge. These are distributed as a means of cell potential control by ingenious action of a special protein pump called the Na+/K+ or sodium potassium pump. There are a large number of these embedded in the cell membrane.

You will remember, from our discussion on replication, that a new cell is created by the process of replication, mitosis and cytokinesis. Such a new cell is already endowed with its correct membrane voltage potential. However, during the cell's normal workings, substances are being manufactured, waste by-products are being produced and substances are being transported in and out of the cell. This ongoing activity causes various changes in concentrations of ions both in and outside the cell which threaten the parameters of that particular cell's voltage potential. The sodium potassium pump works more or less constantly to maintain the cell's voltage potential within its optimal working parameters. Again, this pump uses the energy released in the reduction of adenosine triphosphate, (ATP) to adenosine diphosphate, (ADP) to power its function. The sodium potassium pump uses as much as thirty percent of the average cell's available energy and seventy percent of a nerve cell's. The ATP used to power cellular functions is produced and supplied by the mitochondria, sometimes referred to as the cell's power house.

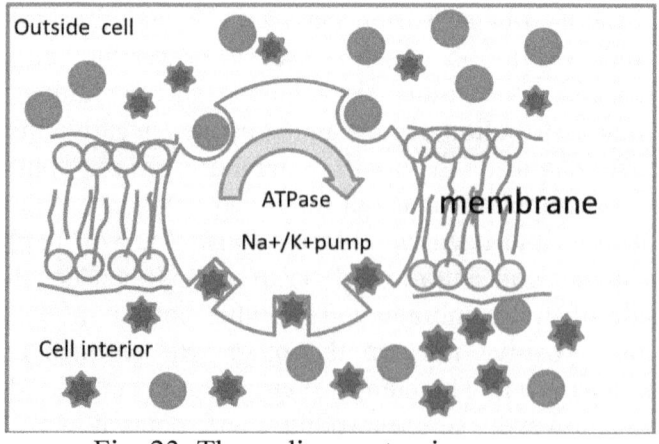

Fig. 23. The sodium potassium pump

Whilst this is not the molecular arrangement or actual mode of operation, the sodium potassium pump is seen here collecting three sodium ions from inside the cell and two potassium ions from outside, ready to rotate and transport to opposite sides of the membrane. This will effectively increase the negativity inside the cell by one ion charge and increase the positivity outside by one ion charge so increasing the net potential across the membrane by two ion charges. The sodium potassium pump can transfer large numbers of ions very rapidly. This is only one of a large number of similarly ingenious, specifically functional mechanisms embedded within and working throughout the cell membrane.

Such is the apparent magic of life; a wonder of wonders indeed. It is worth pondering here for a moment on how life arrived at this mind-bending degree of expertise in chemical and quantum engineering technology. Charles Darwin brilliantly described it as evolution by random mutation and natural selection but what does this actually mean?

Consider our current stage of human technology. This has been arrived at, essentially over the past two hundred years, through a relatively small number of clever inventors, mostly strapped for cash, and some specifically directional research and development facilities numbering about a handful and growing to less than ten thousand over that two hundred years.

Now consider the mindless life process. Even if you want to consider human invention going back some two and a half million years where it is arguable whether they were inventions or discoveries. This is an infinitesimal amount of time compared to the age of the life process at some three and a half thousand million years. Also, we have seen how life evolves by means of a low error incidence in replication. But every living cell is subject to this same small degree of error and there are a hundred trillion cells in your body alone. There have been as many living cells in our world as there are grains of sand in all of its beaches and deserts put together; countless quadrillions of quadrillions of mindless, directionless research and development facilities working over eons of time and with their outputs optimised through natural selection. This is why I refer to evolution as perpetual, multidirectional experimentation in structural form and function. And ponder for a moment on the incredible, technological heights this has reached. We stand in awe of our own technological heights such as rocket science, television, GPS, mobile phones and so on. But think of that which we take completely for granted every day, your ability to see. That the physical reality of the world around you in terms of dimension, shape, colour, texture, movement and perspective is clearly presented as a conscious perception within your mind. This is indeed a wonder of wonders as are your other senses and the moods they induce. You no doubt have experienced those moments when a smell or a taste or a colour or the sound of a piece of music has induced that strange sensation, we call nostalgia; a brief, inexplicable moment of exhilarating romance. Such are the heights and wonder of life's technologies. Little wonder you are described as the most highly ordered and complex entity in the known universe.

Chapter 6 - About the Origin of Life

You now have, at least, a vague picture of some of the gross activities going on within a living cell. You have seen how the system is overwhelmingly complex, utterly logical, so intricately interdependent, so expansively rich in co-related, meaningful information, so ingeniously manipulative, so comprehensively derivative of means and self-perpetuating, so encompassing of our planet and, which as you read, contemplates itself and whose complete elucidation still frustrates the highest echelons of modern science. You can therefore now address in your own mind the eternal questions; how, when, where and why could such a wondrous system come to be? In other words, "what is life?"

You will already be aware of the three classical approaches to the question. One, Creationism which has been in the mind of man since he was first able to think and which was later personified in the biblical stories.

Two: Vitalism which recognises and appreciates the fundamental and categorical difference between the living and the non- living. It attributes properties to life which are not explainable in terms of our present understanding of matter. Some vitalists support the idea that these properties exert a teleological influence on life and that life is therefore a progressive system.

Three: Abiogenesis which is a modified version of the old spontaneous generation idea. It is accepted by modern science as the evolving continuance of a singular, purely natural and accidental event in the distant past. It sees life as an ongoing, non-progressive and purely incremental series of such extremely fortuitous accidents. The first two approaches are, to some extent, interrelated so we shall leave them for the moment and examine, as best we can, the current scientific view. This can only be, in terms of how it might have happened, highly speculative.

Abiogenesis

Firstly, there is, as yet, no scientific theory on the origin of life. There are however, myriad hypotheses or, in other words, educated guesses, albeit, pretty well educated. These include a considerable number of both agreeing and conflicting ideas. In our present era, they are subject to frequent modification. This is due to new information being derived from advancing commercial technologies in deep earth drilling and deep sea robotic, visual surveys and specimen procurement. These reveal that life encompasses an astonishingly wider sphere of our planet and environmental conditions than was ever previously imagined. Until fairly recently it was thought that subsurface life extended to just a few meters under the surface of the earth. We now know that it extends to a depth of at least seven kilometres; some four and a half miles. Also, organisms live in and around deep sea volcanic and hydrothermal vents at temperatures and within environments that would kill most surface bacteria. The chemistries of deep dwelling organisms, both sea and land, considerably widens the sphere of possibilities regarding how life might have begun on Earth and even on its having come from somewhere else.

Darwin's theory of evolution derived from his studies of fossils together with that of the gross morphologies, behaviourisms, adaptations and consequent speciation of existing life forms. It had little or nothing to do with how life began except to illustrate that it began many hundreds of millions or even billions of years ago. The biblical stories determine a beginning of some seven thousand years ago so Darwin's findings were an astonishingly startling revelation in his time; matched only by the implications of his speciation theory. He knew nothing of nucleic acids or the fundamental workings of life and made little reference to how it might have begun. He did briefly suggest the possibility of a "Warm little pool" by the side of a river estuary being constantly nourished by myriad minerals and other substances washed down from eroding rocks and subject to energy inputs from the sun and lightning flashes being a likely situation in which, given long enough, some elementary form of life might establish itself. This became a classical notion and profoundly influenced many eminent scientists including the Russian scientist Alexander Oparin and a British contemporary, J.B.S. Haldane. It was also the basis on which the Miller Urey experiment described earlier was carried out. What I failed to mention in describing that experiment however, is that it was seriously frigged in favour of its desired outcome. In nature, the energy inputs which are capable of initiating synthesis of basic organic compounds are just as efficient in destroying them. This is a difficult problem with these kinds of ideas. In the Miller Urey experiment, laboratory conditions were frigged to circumvent this. Any organic compounds formed in the experiment were siphoned off as they were formed and stored in a special device called a "Cold trap". This then protected them from the conditions that formed them and which would just have quickly destroyed them. No such device exists in nature. However, the Miller Urey experiment was still ingenious and significant. It did have the net value of demonstrating that the building blocks of life could form through natural processes even though, in Darwin's scenario, they would not survive. All that is needed is a different scenario and this forms the crux of many current speculations on the beginning of life.

One of the great questions within the scientific community in speculating on the beginning of life is, did its information technology i.e. the gene come first followed by its hardware i.e. the protein or was it the other way around? This, in our computer literate society, may seem a rather ridiculous question. Who on earth would attempt to build a computer without first knowing the way in which it would be programmed to function? It is not as ridiculous as it first seems however. Our computer age was preceded by a wide range of functionally predetermined, mechanical calculating machines. These were used mainly in commerce and included comptometers, the McClure multiplier and a wide range of adding, subtracting and tabulating machines all of which, in modern jargon, would be described as hardware.

Independent of Alan Turing's secret project at Bletchley Park, these evolved during the early nineteen fifties through the vacuum tube or glass enshrouded valve in conjunction with diodes and later, the early transistor to an in between stage with punched card programming; the beginning of information technology. This evolved further to our current stage of the microchip with integrated circuitry and the development of fully programmed information systems. However, there are those who would argue, quite legitimately, that the hardware machines mentioned were preceded by information technology existing as intellectual notions in the minds of those who designed them. In fact, to my unqualified mind, questioning which came first in terms of life's information or hardware technologies is a kind of romanticism and is hardly a valid question. Life's information aspect lies in the physical format of the DNA molecule which, in itself, is just as much hardware as is a protein. Information of any kind exists only in the capacity of the conscious human mind to perceive it as such. Yet, there still remains the bewitching chicken and egg problem. The information aspect of life requires the action of specifically functioning proteins to both form and utilise its representative, molecular structure whilst, at the same time, the structure and functionality of a protein depends upon utilisation of the information represented. It is rather interesting that modern science should consider life in terms of hardware and software. It is, at least, a hint of metaphysical
136

interference which it otherwise, and in every aspect, ardently denies.

During the early and mid-nineteen hundreds, the Russian scientist Alexander Oparin and a British contemporary, J.B.S. Haldane, envisaged concurrently, whilst speculating on the beginning of life, what science calls a protein (Or hardware first) kind of evolution. It became one of the most widely accepted hypotheses in the scientific community and continues to have much respect to this day. Their hypothesis centres on a phenomenon called Coacervates.

Explaining what coacervates are, even though they are a common natural phenomenon, is not easy in layman's terms. This is because, whilst in the foregoing description of the central dogma of biology we were able to consider a series of gross activities and thus avoid the immense technical detail, explaining coacervates would require going into the nitty-gritty of chemistry. We shall avoid this therefore and opt for a very simple demonstration which, with very little effort, you can carry out for yourself.

Find a soft earth patch in your garden and, with a trowel or similar instrument, dig a small hole to a depth of about five hundred millimetres or eighteen to twenty inches. In most cases, you will find that you have reached a layer of smooth, evenly coloured clay. Take out a piece of clay and cut out one cubic centimetre. Put the piece of clay into a clear jar or glass together with one and a half tablespoonfuls of water. Work the clay with a blunt piece of stick until it is completely and smoothly dissolved in the water. Now take a shallow saucer and pour into it a few millimetres deep of white spirit. Take half a small egg- spoonful of the dissolved clay and carefully drop it from a height of about one centimetre into the white spirit. You will see three things. One is a clear demonstration of substance immiscibility; one is the natural uniformity of that effect in the clay solution forming a near perfect circle on the bottom of the saucer and the other is the perfect clarity of definitive boundary between the clay solution and the white spirit. This is not what one would call a coacervate and, in fact,

137

might be called the opposite of a coacervate, but it is a macroscopic demonstration of the main forces at play in coacervate formation. A real coacervate is a colloidal globule of organic lipid substance in water which is self-organising on the same basis as that in our little experiment. They measure only 1 to 100 micrometres (Thousandths of one millimetre) in diameter and exhibit osmotic properties and even some degree of organic molecule manipulation across their boundaries.

Oparin and Haldane were competent and devoted scientists and were consumed in their efforts to find some answers to the strange workings and beginnings of life. They saw a striking resemblance in the natural, self-organizing, globular and permeable characteristics of coacervates and that of living cell membranes. They imagined some kind of coacervate enclosing an environment wherein some kind of elementary protein might be formed. Since such a protein would be a product of that environment, it would be conceivable that it might react to its environment in such a way as to further enhance it in terms of protein formation and function, so introducing an evolutionary life trend. Also, coacervates can meet and combine and, dependent on size, can suffer mechanical stresses which cause them to break or divide. Oparin and Haldane saw this simple mechanism as a possible starting point leading eventually to a primitive form of cytokinesis.

Oparin and Haldane were proponents of the now generally accepted theory of primordial Earth having an oxygen free atmosphere. This is because oxygen inhibits the synthesis of organic compounds. With no oxygen in the atmosphere, there would be no ozone layer so the surface waters would be bathed in high energy ultraviolet radiation. This would power the synthesis of organic materials, now known as the Haldane soup. This could collect within the confines of a coacervate where it would be, at least, partially protected from the ultraviolet radiation. The Earth was still an extremely violent place at the time life first appeared. As well as volcanic and tectonic activity, there was extremely wild interaction between seas and shores. Different types of coacervates can be formed enclosing different clays resulting in different chemistries. Violent Ocean to shoreline activity could produce millions or billions of coacervates with a range of different internal chemistries so providing a wide sphere of chance for a beginning of life.

According to the Oparin Haldane model, the first life would have been in the form of simple anaerobic heterotrophs. Anaerobic means without oxygen and heterotrophs are organisms which scavenge and consume organic materials as food and convert them into energy and tissue. In the early Earth and in some microbes today, this metabolism would be by fermentation. Fermentation is the breaking down of organic substances such as sugars or starches into an alcohol or an acid through the action of specialised enzymes. During the process, electrons are released and re-appropriated to reorganise the original molecular structure of the nutrient. It produces two molecules of ATP per molecule of sugar which the organism can then use to power its metabolism. However, this is describing a rather advanced life process so the Oparin Haldane model is a broad concept covering millions to hundreds of millions of years of evolution. Some scientists see this scheme as being beset with serious difficulties not the least of which is that a coacervate may be too fragile to sustain any particularly conducive environment for anything like an evolutionary time span. Also, it is difficult to construe any means whereby simple mechanical stress might form a basis for the development of the complex, interdependent, chemical, molecular machinery involved in cell mitosis and cytokinesis.

Let us assume for a moment that a coacervate did, by some means, circumvent the necessity and continuity of ribosome activity and augment a linkage of amino acids in absence of any genetic influence. In order for it to instigate even the most elemental life function, it would have to have a meaningfulness in those terms; one meaningfulness in an almost infinite number of other meaningless contingencies. It would be like finding Bobby's atom. There are a number of scientific research programmes being carried on at present involving protein manipulation. These produce some alterations in protein structure and function and the purpose of the science is to find means whereby useful, therapeutic nano machines can be constructed. There has already been some amount of success in this. However, the basis of this science lies primarily in gene manipulation and subsequent expression. Any meaningfulness in such proteins therefore is an extension of the same meaningfulness already pre-established by and inherent in the life process. This is a problem with origin of life scenarios that is much exploited by opponents of evolutionary theory whose argument is something like what follows.

Consider a functional protein forming by chance and in absence of any genetic influence. We have no way of knowing what the first protein might have been like in terms of make-up or function. However, in order for it to have been a precursor to life, it had to be capable of sustainment by interaction with its environment and have the capacity to replicate or be an active component in an already replicative system which imprinted all of its attributes, including replication, in its progeny. We have to conclude therefore, that even such a first protein would be endowed with a very considerable degree of complexity. A typical protein contains from two to three hundred amino acids and there are some which contain many thousands. We are going to consider one of the simplest proteins. It is one we already came across in the gene expression mechanism, ribonuclease. This numerically simple protein is comprised of one hundred and twenty-four amino acids and includes seventeen of the available twenty different kinds. These occur in different frequencies along the length of its chain. Ribonuclease cannot replicate on its own and its function is merely to act as a disruptive and degrading influence on RNAs; except for its selectivity, not greatly different from that of washing up liquid on the residues on your dinner plate.

The chance of the first amino acid of ribonuclease, which is lysine, arriving in place is 1 in 17. The second, glutamic acid, is 1 in 289, the fourth, which is three alanines, is 1 in 83,521 and the sixth, which is phenylalanine, is one in twenty-four million. By the time we get to the seventeenth amino acid, which is valine, the chance is now one in ten to the power of 152 and even this depends on a constant supply and mixing of all these ingredients.

You will remember that the number of atoms in our entire universe is one by ten to the power of eighty. Ribonuclease, or a similar sized protein, forming by pure chance therefore, would be like finding Bobby's atom, not just in the entire mass of our planet or even that of the solar system or the Milky Way galaxy or even in our entire universe. It would be like finding Bobby's atom lost somewhere in a universe of universes where it would be miraculous to find a particular universe far less a particular single atom. It would be a miracle in the most literate sense of the word. Also, a first replicating protein would have to have been much more complex than ribonuclease.

The foregoing seems a rather disheartening scenario in terms of theorising an origin of life. At first sight, it seems extremely compelling in favour of creationists in support of life having originated by omnipotent design. However, it is a trick argument, aimed at the uninitiated, and treats amino acids exactly like our hundred pennies. In our hundred pennies, it is worth noting that, in terms of relationship between one penny and another, the pennies are quite inert. The pennies have no interacting properties or influences whatsoever between one another that might influence the order in which they fall. On the other hand, we have seen, from our brief consideration of life chemistry, that atoms and molecules can have many interacting influences on one another. We have seen, for example, how readily carbon reacts with hydrogen and the almost infinite sphere of chemistry arising from substituted hydrocarbons. These relationships bring the complexities of life chemistries out of the next to miracle chance of our hundred pennies and into a realm of feasible probability. Even so, the hardest of evolutionists, like myself, have to accept that the degree of "Feasible probability" is close to being miraculous.

Science attributes certain characteristics in its description of life. For example, an organism must be capable of replication. It must be capable of interacting with its environment in terms of sustenance and waste excretion, capable of isolating its own internal environment from the environmental medium in which it lives, capable of maintaining its own internal environment against changes in its external environment, it can grow and it evolves. There are no doubt other characteristics attributable to life. However, in my studies for the purpose of writing this work, I was made aware of a very important characteristic of life which I do not see mentioned anywhere, perhaps because it is so obvious. That is, that life exists within a sphere of total functional interdependence. What this means is, that your small toe toe nail has a degree of functional relationship with every other part of your body including your brain. This degree of interdependence varies throughout your body to a state of totality with your major organs. On this basis therefore, it would seem that any origin of life hypothesis which chronologically separates the software and hardware aspects of life is likely to be wide of the actual truth. It seems much more likely that these aspects would have to occur and develop concurrently in a functionally interdependent manner.

Whilst the Oparin Haldane model still commands a considerable degree of respect within the scientific community, there are now a considerable number of alternative scenarios. Unfortunately, any explanation of these would only be meaningful within a good understanding of chemistry. However, we can briefly consider one or two of the more attractive.

We have already considered what a catalyst is as a facilitator and speed enhancer of chemical reactions. Most modern, realistic hypotheses on the origin of life associate it with the natural catalyst phenomenon. Whilst particular catalysts are normally associated with particular reactions, there are many catalysts existing outside the sphere of life. During the hellish turmoil of Earth's beginning and subsequent early existence, there must have been a considerable number of different and encapsulated environments.

142

Some of these, in chemical terms, may have been quite exotic. If, within one of these, a catalyst we shall call A persuaded the formation of a substance we shall call B and B was a catalyst forming a substance C which catalysed the formation of the initial substance A, you now have, depending on resources, a system which is interdependent, dynamic and repeating. Also, such a system is a physical function which contains information in terms of its molecular configurations and chemical affinities. This kind of system is known as a catalytic cycle or autocatalysis. The only product of such a system however, is its own persistence by constant repetition of the cycle so long as it is within an environment that satisfies its energy and resource requirement. It might be considered as the chance coming together of a few fragments of information which, whilst not being alive as such, encompasses a sphere of dynamics reminiscent of those involved in the life process. The only problem with this scenario is the requirement for a sufficient amount of the right catalysts and associated compounds to be present in the same place at the same time. However, in terms of our chance formulation, this only requires a bundle of pennies. It doesn't require them to be in order. The order in which such a system self-organises, like our coacervates, is already specified in terms of natural affinities between its chemical components. Furthermore, the product of such a catalytic cycle can, as a whole, act as a catalyst persuading the formation of another compound outside its own catalytic cycle. If this new compound also happens to be a catalyst, then it may become an active component in another separate and different catalytic cycle. Such a system will then preserve an element of information through repetition and produce a profusion of such entities. If one of these should act as a catalyst persuading the formation of our original catalyst A, we then have a system which is replicative rather than just repetitive. Such a system is called an autocatalytic hyper cycle or catalytic set. The workings of the DNA molecule have been demonstrated to be a catalytic set. Most of the scientific community now leans heavily towards autocatalysis being the fundamental chemical precursor to life.

Other hypotheses are centred on chemistries existing in and around deep-sea hydrothermal vents where whole ecologies are found to exist. Without going into the actual chemistries of these ideas, they would indicate that the first living things were autotroughs rather than heterotroughs. Autotroughs are organisms that make their own food from a lower but more common level of substrates rather than foraging for organic materials as in the Oparin Haldane scenario. These hypotheses have the great advantage that autotroughs stand to this day as the main food base upon which all further evolved heterotroughs depend for their survival. Autotroughs are comprised of chemotroughs and phototroughs. Plants and green algae are phototroughs. They derive their energy from sunlight by the process of photosynthesis.

A typical example of a chemotroughic origin of life is one based on the chemistry of undersea, alkaline, hydrothermal vents which are not volcanic. They are formed by seawater penetrating through cracks in the rock of the Earth's mantle. These rocks are rich in electrons and the water reacts with the rock in such a way as to cause it to expand and split in a manner similar to the modern technology of fracking. This allows more water to percolate so producing a continuous reaction. The reaction produces alkaline fluids which are proton poor. It also produces heat, causing a thermal gradient which drives the alkaline fluids up to the ocean floor. When these come into contact with the cooler ocean water, the minerals precipitate out and form the crusted chimney-like structures. Since there was little or no oxygen on the early Earth, the oceans would be rich in dissolved iron and carbon dioxide so the ocean would be acidic and therefore rich in protons. This means a proton gradient exists between the alkaline chimney and the ocean water. This would provide sufficient energy to drive chemosynthesis within the porous chemistry of the vent crust. This idea assumes that fine separations between cavities in the vent crust might act as primitive membranes and may have initiated an

144

autocatalytic scenario. This, in turn, may have evolved into a simple synthesizing and replicating organism.

An even more attractive hypothesis is one called the RNA World Hypothesis. Probably, the earliest proponent of this idea was Francis Crick. However, it wasn't until nineteen eighty-six that it was called the "RNA World" by the Nobel Laureate Walter Gilbert and it is from this hypothesis that the word "Ribozyme" emerged. You may remember, from the Central Dogma, that a ribozyme is an RNA molecule which has the capacity to act as an enzyme in a manner similar to that of some proteins. A great attraction of the RNA world hypothesis is that it describes a conjunctive origin of both the information and hardware aspects of life. Like other hypotheses it has difficulty with the originating chemistry but adheres to the idea that life emerged from a form of autocatalysis. The idea predated the current knowledge and understanding of under-sea hydrothermal vents and was proposed as an enhancement of the Oparin Haldane scenario. In the light of newer, modern understanding it would seem that elements of all or most of the varying scientific hypothesis might combine in providing a more wholesome scenario. For example, it is difficult to imagine how fine separations in the porosity of the crusted material of a hydrothermal vent chimney might act as a membrane. It is also difficult to imagine a coacervate persisting for long enough to allow accumulation of any complex molecular activity. However, it is feasible for a large number of coacervates to exist in the locale of a vent chimney and for them to enclose a widely varying variety of clay solutions arising from the vent activity. It is also feasible that coacervates could become entrapped within the porosity of the chimney crust. This may then provide the coacervate with protection and, at the same time, provide sufficient exposure for its membranous activity. The coacervate might then augment a chemotropic rather than heterotrophic origin of life.

There are a considerable number of other hypotheses on the origin of life but, unfortunately, they all have similar endings. These amount to various ways of saying "This might have eventually led to a first primitive, self-replicating, living organism." So, despite modern science's belief that life emerged from non- living matter, the statement made by Louis Pasteur in eighteen fifty-nine that "Life cannot emerge from non-life" still stands... Modern science is, as yet, unable to concoct any circumstance in which life self-initiates without an already living ingredient. This however, is by no means proof that it didn't happen. My personal belief is that it did and, whether or not it initiated on our planet, that it probably initiated in autotrophic form through some combination of the scenarios we have considered.

Notwithstanding the foregoing, there still remains a broken bridge in understanding how even complex autocatalytic chemistry evolved into a living organism. The roots of all life as we know it lie in the nucleic acids and there is, as yet, no scientific consensus on how these came about. They are necessary for life yet seem also to be a product of life; the bottom of the barrel in the frustrating chicken and egg dilemma. However, we have seen that life thrives through natural chemistry. We have also seen that natural chemistry has the capacity for more or less infinite ramification which, in the life scenario, has been expressed as multidirectional experimentation in structural form and function for at least four thousand million years. Science has had a reasonable capacity to investigate the culminate state of these immensely accrued complexities for only one hundred years or less. We have to admire science for the considerable degree of understanding it has achieved in that short time and which is more than sufficient for us to know that life, however it happened, is a purely natural process. Life sets its own stage with a mesmerizing magic but it is not a miracle.

In considering scenarios for the origin of life, one must bear in mind that the event occurred at least four thousand million years ago. There may have been circumstances existing then which, however ingenious the concoction and however sophisticated the laboratory, are just not reproducible now. Consider, for example, if atomic radiation was involved in the initial circumstance. This is not as outlandish as it might seem. During the disassembly of the Chernobyl nuclear power plant which suffered melt-down in nineteen eighty-six, steel plates were being removed from a highly radioactive area. Upon removal, it was discovered that the undersides of the plates were covered in a fungus which was surviving on radioactive energy. If radioactivity played a part in the origin of life, it may have involved a particular half-life state of one or more elements. If that was the case, science may never be able to re-create the circumstances because that half-life state has now passed on our planet and will never again recur. This is only one example of what might be an arm's length of non-reproducible circumstances; circumstances that might even be of extra-terrestrial origin.

Some scientists believe that life is an inevitable occurrence in our universe and some believe that when conditions are right for it, it happens very quickly. From what we have seen of chemistry, there may be a case for its inevitability. But since there is no knowledge of how it actually happens, there is no way of knowing how quickly it might happen. At the same time, cellular life can be traced back to within only half a billion years of the formation of the earth. This first recognisable cellular life is still highly complex and therefore represents a considerable period of evolution. This means that, if life began on Earth, its beginning, in terms of evolutionary time spans, was almost coincident with the formation of the Earth which, in turn, means it did indeed happen very quickly. Alternatively, did it begin somewhere else? If it takes a long time to happen, it had twice as long to happen before the formation of the Earth than it has had since.

My personal leaning is heavily in the direction that, whilst life most likely originated by some combination of the scenarios mentioned, it did not originate on earth. Archaic cells can be traced to only half a billion years from Earth's formation. This means that if life originated here, it's evolution from a basic chemistry to a living cell containing all the essential and incredibly complex machinery, amounting to a principle of life, took place during a

147

relatively short time and when our planet was most inhospitable to life. Life evolution is a slow process, even after establishment of the principle. It must have been incredibly slower before such establishment, even though there are scientists who glibly say it might happen very quickly. Life evolves through random mutation so it must have originated through a course of random change. Within such a system, changes which are disadvantageous to the system must outnumber those which are advantageous by millions or billions fold. Every such disadvantageous change is a setback which must be compensated for and therefore acts as a multiplier in terms of evolutionary time. It is only after the principle has been established that it is endowed with some degree of selection in this respect. I am strongly inclined therefore, to the idea that life originated somewhere else, probably over a period of billions of years and was deposited on our planet during and after its formation. This leads to a consideration of that forming period.

Chapter 7 - Life, The solar System and Prospect of ET

According to science, our Solar System began to form some four and a half to five thousand million years ago through a process known as the Nebular Hypothesis. It started with the gravitational collapse of part of a huge cloud of interstellar gas and dust. Most of this material, some ninety-nine-point eight percent, was gravitationally oriented into a central mass, forming the Sun. The remainder became subject to a number of forces including gas pressure, gravity, magnetic fields and intense solar winds generated by the forming Sun, inertia and angular momentum so that it eventually formed a planer, protoplanetary disc. The planets, including our Earth, were formed from this by the method known as accretion. Accretion is the coming together and consolidating of gas and or dust particles influenced by their mutual gravitational attractions. This generates areas of higher density dust clouds which attract one another and produce even higher densities. This process continues until bodies become sufficiently massive to exhibit a reasonable degree of gravitational coherence and form planetesimals. These bodies are estimated to have been in the order of one to ten kilometres in diameter and underwent further series of collisions and mergers to form protoplanets. The protoplanets then further collided and merged to form the planets and other orbital debris as we see them today.

This scenario was first proposed by the Swedish scientist and philosopher Emanuel Swedenborg in the eighteenth century. It is interesting to note here that Swedenborg was obviously a person of high intellect. Yet, on several occasions, he did exhibit psychic powers, including predicting the exact date of his death, and, in his writings, claims to have visited the gateway to Heaven and Hell and to have conversed with angels. However, his Nebular Hypothesis has fallen in and out of favour over the years but has gained considerable support from recent observations, by modern sophisticated equipment, of stars surrounded by protoplanetary discs. The nebular Hypothesis is now generally accepted by science. The detail of this scenario however, is not without serious problems.

Consider a sewing needle. It is normally made from high quality carbon steel and may also contain some proportion of platinum. In other words, it is comprised of many billions of massive atoms. If you pick one up from the floor, ninety-nine-point nine recurring percent of the energy you use is in the movement of your body. The gravitational effect of the entire mass of planet Earth on the needle is so small and insignificant that you just cannot feel its weight. Now consider a ten-kilometre-wide volume of condensed but gravitationally incoherent dust and gas. Its gravitational effect would be infinitesimal. Also, depending on its temperature, its coherence may be disturbed by natural Browning Motion. Browning Motion is an effect of the kinetic energies of atoms which causes them to randomly bounce off one another. Any collision of such bodies in orbit must surely react to the fierce dynamics of the collision rather than the almost non-existent gravitational influence and disperse rather than accrete. To demonstrate this problem, try going outside and building any three-dimensional geometric shape from the contents of a chicken feather pillow.

There is also the question of molecular cohesion. When people think of accretion, they tend to think not so much of dust but rather of lumps of rock and other mineral and metal objects bombarding
150

and building up planets. The planets in our solar system, as well as accreting from dust and gas, were subject to much bombardment by other debris. But these objects could not possibly accrete in those kinds of sizes. This debris can only have come from other planet sized bodies which were broken up and fragmented by collisions.

Consider two one hundred-millimetre cubes of mild steel. Place these on top of one another and put them into a large, powerful hydraulic press. Apply say three hundred tons of pressure. Now release the pressure and remove the blocks of steel from the press. You will find you still have two separate pieces of metal. The immense pressure of three hundred tons was insufficient to bring the two pieces of metal close enough together to achieve molecular cohesion. In order to achieve molecular cohesion, the metal needs to be heated to a gelling or liquefying temperature. The same applies to other minerals and rocks. To achieve this by gravitational pressure requires a body of sufficient mass to initiate geothermal activity, something like twenty percent Earth's mass.

I see the nebular hypothesis as a perfectly viable scenario for star formation in the early universe when gas clouds were very much denser than they are now and there were no such things as planets. Later, the dust in the gas clouds can only have come from supernovae explosions and it is reasonable to assume that such explosions also produce a significant amount of coherent, solid debris. I lean towards the idea that gas and dust clouds are gravitationally seeded by such debris which initially promotes the gravitational accretion of stars and planets against the disruptive forces of momentum, inertia and temperature. Some of this debris might be comprised of planets or parts of planets that have been ejected into interstellar space by the supernovae or other nearby exploding stars. Some of it may have existed as planets for as long as two to four billion years before our Sun was born and may have already been endowed with life at the time our Solar System was being formed. This would account for what seems to be a coincident occurrence of our Earth and life. It also means we should expect to find life elsewhere.

151

So, what of ET.? Where are all the Klingons, the Kardashians, the Romulans and Feringie? Where are Edgar Rice Burroughs' huge banths and maidens of Mars? We are surrounded by hundreds of millions of billions of stars and recent observations through highly sophisticated instrumentation indicate that many millions of them support planetary systems. These were formed in the same manner as our Solar System. They are comprised of the exact same elements which promote the exact same chemistries and, amongst such profusion, there must be many existing in Goldilocks zones. Yet, since Nikola Tesla in eighteen ninety-nine and through the highly sophisticated technologies of modern S.E.T.I. (Search for extra-terrestrial intelligence), not a single morsel of intelligent communication has reached us from space. With the use of huge radio telescopes and spectrum analysers, S.E.T.I. has had the capacity to simultaneously search hundreds of millions of narrow band channels yet the depressing silence persists. So why should this be? Is it simply a question of differing communication technologies? I hardly think so. The electromagnetic spectrum is tied up in universal constants and we have already developed a wide range of signal detectors. Even if an intelligence had developed a more advanced communications technology, say in the realm of lasers or particle entanglement, there would still be others like us and not that far advanced. So, are we the only intelligent beings existing in our universe?

When considering this question, one must first consider our interpretation of what we call intelligence. If you have ever watched Star Trek or watched or read anything else which portrays imaginary aliens, you will have noted that they are always presented in varying degrees of uglification of the humanoid form. It would be stupid of a writer or a director to portray a dog or a horse or an elephant or crocodile wielding a weapon or manipulating the controls of a space ship. Try to imagine any animal outside the sphere of primates trying to deal with the simple task of fastening a shirt button. There is a deeper reason for this. Your intelligence is not so much a function of your brain as it is of the entire electro-biochemical and mechanistic configuration of your body as a whole; the way in which your vision works. The ways in which your torso, your shoulders, your arms and your hands articulate together with your facility for spatial mobility are as essential a part of your brain function as is the brain tissue itself. This is also true of the generality of your nervous and sensory systems. The degree of interdependence in the functionality of any living system determines that your body and your brain evolve as one. Each is a functional part of the other. Our intelligence therefore, is largely a product of our overall biological configuration. All of this can be brought into stark perspective by asking one's self the simple question, is a dolphin or a dog or a chimpanzee ever likely to feel frustrated because it is unable to use a sewing needle? (Some recent scientific research would indicate that some animals like dolphins, octopuses, crows and chimpanzees have a spark of imaginative faculty) Our comprehension of the universe is a product of our particular biological configuration and, in evolutionary terms, this is literally one random fall of the pennies. In this respect, we may be unique even in the unimaginable vastness of our universe. The Drake equation may be infantile in its naivety. Also, you can see from this, that even if such uniqueness were provable, it would still not imply creationism. We shall briefly consider this profound question of ET later when we shall see that our universal solitude may be real and for good reason.

There is also the question of course, are there different kinds of intelligence? Different animals exhibit different degrees of intelligence beyond what we can describe as instinctive behaviourism. But their intelligence has followed the exact same evolutionary track as our own and differs only in degree but not in kind. If we were to encounter an intelligence which was different in kind to our own would we recognise it as such? Science seems to be convinced that we would recognise any intelligence through its understanding of the universality of physical laws and rationality of mathematics. But that would only be true of an intelligence emanating from biological mechanistics similar to our own and following the same evolutionary and technological course. We have an example of that dilemma to hand. As discussed earlier, science can eloquently describe what life does but not what it is. Just as our intelligence pushes inert substances like sand, gravel, stone, metal ores and so on into the vibrant technologies of buildings, engines, cars, aircraft, television, mobile phones and computers, life contrives to push chemistries in the complete opposite direction of every other universal tendency. It achieves the highest degrees of physical complexity in the known universe yet seems to rely on a form of chance rather than what we see as logical connectiveness. By random application, life achieves the absolute in multidirectional experimentation in structural form and function, carried out over eons of time, aligned to order by the universal laws of physics and brilliantly described by Darwin as evolution by natural selection. Is life the essence of a different kind of intelligence? Might this be the underlying fundamentality of what science sees as the airy-fairy concept of Vitalism? Is Richard Dawkins' Blind Watchmaker really so blind?

This is not to say there isn't life in other parts of the universe. In the chemistries we have considered, there is nothing to indicate that they should be unique to Earth. On the contrary, together with recent observational confirmation of the Nebular Hypothesis, the commonality of planetary systems and our understanding of the universality of physical laws, they demand our acceptance of life being reasonably proliferate in the wider universe. Life might not be a miracle on Earth but it would be miraculously inexplicable if it were confined to Earth. However, intelligent life may be quite something else. Whether life was deposited on Earth close to the time of its formation or actually began here, its beginning on Earth was either in single celled microbial form or something which led to the assemblage of a primitive, single celled microbe. It remained in this relatively simple form for some three and a half thousand million years and might well have remained so to this day except

154

for a course of some rather freakish falls of the pennies. There is no natural law determining that life inevitably leads to intelligence.

On the contrary, and without going into the mathematical and technical details, there is an energy relationship between the information content of a living cell i.e. its genome, and the amount of chemical synthesis and other metabolism specified by the genome. Throughout most of the time life has existed on our planet, this relationship has been such that it was not thermodynamically favourable or even possible for an organism to be large and or highly complex. This effectively imposed a thermodynamic clamp limiting the size and complexity of organisms. It restricted life on our planet to a single celled, microbial realm of being for some ninety percent of the current lifetime of our planet.

This is probably why science is having so much difficulty in discovering past life on Mars. It is very likely that life did exist on Mars. However, it is most unlikely to have ever evolved beyond the single celled, microbial realm. It may have been confined to such diminutive size and delicacy of substance that it would leave no detectable trace that would survive geological time spans.

So how did we ever arrive at two-hundred-ton whales, eighty-ton dinosaurs and twelve-ton elephants? How did we ever arrive at the, as yet, unfathomable complexity of human intelligence?

Consider again, you're throwing your one hundred numbered pennies into the air from your shovel. The pennies will fall. This is in accordance with physical laws already well understood. And, when they fall, they will fall in some kind of pattern which, in terms of the goal you are seeking, will be completely random. What is interesting however is that that particular random fall is endowed with the exact same degree of uniqueness as the particular fall you are seeking. It was subject to the exact same degree of chance and, no matter how often you throw the pennies; it is extremely unlikely ever to repeat. Also, so far as science can determine, life has only occurred once on our planet. All other life is descendent from that singular instance. This same principle

155

applies to mutation and the random direction of evolution. This is self-evident from the variation in life you see around you. Nature throws the pennies constantly, instant to instant, over eons of time and the DNA molecule has acquired the capacity to recognise, select and store potential utilities from the ongoing random falls. This is the essence of multidirectional experimentation in structural form and function.

When life first appeared on our planet, its first emergence was in single celled, archaic or bacterial form and was almost certainly chemotrophic. It reproduced asexually by binary fission. Binary fission is equal appropriation of the cell's contents into two hemispheres of the outer membrane and division of the cell into two new daughter cells. These, in turn, will reproduce and produce four new daughter cells. You can see from this that, the survival rate only needs to minimally exceed the death rate, to cause an exponential increase in microbe numbers. The environment in which the first microbe appeared and survived must have been highly conducive to the microbe's needs so it would soon grow into a colony of considerable number in that area. This happened when our planet was little more than half a billion years old, so the planet was still a very violent place. Powerful gales, torrential rains, changes in temperature, eddies, currents and enormous tides would disturb the colony. It would be broken up and distributed over ever widening areas. Fragments of the original colony would continue to reproduce and multiply but, in order to survive, would have to adapt to their various changing environments. But how does a mindless little microbe adapt?

Since this work is intended for lay reading, and, since adaptation is another one of the things the anti-evolutionist pounces on when trying to recruit the uninitiated to his or her way of thinking, it is worth some consideration here.

You have a sensual and intellectual awareness of your environment and can decide to do things when your environment changes. In colder weather or if you move to a different climate you change your clothing. When swimming in cold water, you wear a wet suit. You wear appropriate footwear for different terrains etc. The anti-evolutionist will argue that microbes have no intellectual capacity to recognise any need for change nor are they equipped to affect any changes in their internal workings or overall morphologies in order to accommodate changes in their environments or otherwise competitive situations. All of this, according to them, is affected by the intelligent will of God. At first sight, it seems a rather compelling argument.

Understanding this question lies in understanding that organisms don't really adapt; well they don't and they do. You will remember, from the Central Dogma, that whilst the DNA molecule is about the most reliable information store ever created, it is subject to a certain error incidence due to copy failures and other mutational interference. Now consider our original colony of microbes and the fact that an essential quality for life is the capacity to replicate. Also, replication itself implies inheritance. When an organism replicates (Or reproduces), all of its attributes are inherited by its progeny. Even though the DNA molecule is extremely reliable, there will always be some microbes in our colony whose genes are changed by the means already mentioned. The number of microbes affected in this way will depend, to some extent, on the number of microbes in the colony. This may be millions or billions of which some will be mutationally affected. Of these, some may perish in their existing environment. Some may be left struggling to survive, some may remain unaffected and some may enjoy an enhanced compatibility with their existing environment. If the environment is now changed either by the colony or part of it being moved by other external forces or by a change in temperature or salinity or acidity etc., most of the colony is likely to perish. But amongst those whose genes have been altered by mutation, there is likely to be a few who are now just as suited to the new environment as the main colony was to the previous one. These will survive and reproduce in the new environment and the colony will grow again. But whilst these new microbes are the same species as the original colony, they are not the same microbes. The old microbes are all dead but by mutation, replication, inheritance and death, a new seemingly adapted colony now exists. You can see from this, that whilst the original colony did not actually adapt as such, the net outcome is a newly adapted colony. Adaptation is also affected by a number of mathematical factors but the foregoing is a very brief explanation of how it is achieved without need of sense or God.

So, our first colony of microbes begins to gradually spread over ever widening areas of the planet; all the while adapting to changing environments. Also, and by similar means to that of adaptation in conjunction with the optimizing effects of natural selection, is diversifying in form and function into different phyla, species and eventually kingdoms of life. Evolution is in full swing.

It is not clear which came first, archaea or bacteria. They are, for the most part, similar in size and shape. They are distinguished from one another mainly on the basis of some differences in their cell wall compositions and protein synthesizing polymerases. My personal leaning is heavily in favour of the first organisms being archaea. This is because archaea, sometimes referred to as extremophiles, as well as existing in all habitats where bacteria exist, also exist in many extreme habitats where only archaea could live. Wherever life originated therefore, clearly, archaea had, by far, a wider sphere of chance and survivability. It is likely then, that bacteria evolved from archaea and, some three and a half thousand million years ago, bacteria split off from archaea and both forms then went their separate evolutionary ways. It would seem that bacteria, whilst possibly having poorer survivability than archaea, had developed a much greater evolutionary potential in terms of size, diversification and complexity. Some bacteria evolved a rudimentary, non-oxygen producing photosynthesizing mechanism, utilizing the proton gradient principle to power the manufacture of ATP. The proton gradient principle persists in practically all life forms to this day.

Over the next five hundred million years, cyanobacteria evolve. These capture the hydrogens from water molecules and so

158

produce free oxygen as a waste product. At first, this oxygen is captured by oxidizing dissolved iron in the oceans and forming iron ore. Eventually however, all the dissolved iron is oxidized and free oxygen begins to permeate both the oceans and the atmosphere. This spells doom for almost all life on the planet. It takes another two hundred million years for oxygen to reach lethal volume at which time all anaerobic (Oxygen intolerant) microbes, nearly all the life on the planet, is extinguished. This is about the greatest life extinction event in the history of the Earth.

The oxygen in the atmosphere had further devastating effects. The Earth's atmosphere at that time contained a considerable amount of methane, a greenhouse gas. The new atmospheric oxygen reacted with the methane in such a way as to remove its greenhouse effect. This produced one of the most severe and prolonged ice ages in Earth's history; what science refers to as a "Snowball Earth". It lasted for some three hundred million years. Think of the Egyptian pyramids being some five thousand years old or the first modern human being forty thousand years ago. Then try to imagine three hundred million years.

There is no scientific consensus on how this severest of ice ages came to an end. An ice age of such severity tends to be self-perpetuating because it reflects almost all of the Sun's heat away. There are two obvious scenarios however, any one of which may have ended the ice age. One is a significantly large impact event which would have enshrouded the Earth in dust debris and also blackened the surface of the ice. The other is a significant change in the Earth's geothermal activity initiating a period of violent volcanism. This would enshroud the Earth in both debris and gasses which would also reinstate the greenhouse effect. However it happened, it did happen and with just the right degree of severity which allowed the cyanobacteria and some other extremophiles to survive.

After three hundred million years, the ice melted and the Earth returned to some degree of normality. Life entered a new phase of proliferation of chemotrophs and aerobic heterotrophs which were

159

not only oxygen tolerant, but which utilized oxygen in more efficient ways of producing energy. This allowed much microbial life, whilst still being single celled, to become larger and significantly more complex. Many microbes were now large enough to eat or engulf other smaller microbes as well as other organic substances, (The process of phagocytosis) Life was now straining at the seams of the thermodynamic clamp.

Over the next one hundred and fifty million years, a gigantic leap in evolutionary progression, (If I may make so bold as to couple these words in that way), occurs. May I refer you back for a moment, to where we were considering the changes in scull bone configuration for the accommodation of brain tissue in the evolution towards modern humans? We saw then, that the DNA molecule seems to have the capacity to anticipate usefulness in certain fragments of information which it then stores in the form of what used to be called junk information. It might be likened to a competent mechanic. If he sees an occasional nut, bolt, washer, split-pin or circlip etc. lying about the workshop floor, he will not pass it. He will pick it up and put it in a junk section of his toolbox. He knows that, one day, a piece of this junk will satisfy an important and immediate need. Most mechanics would feel naked without such a junk collection. The following course of evolutionary events is typical of the interpolation of such previously stored fragmentary information. It might be likened to a person who decides to build a garage for the accommodation of his car. The information relating to the building and functionality of a garage is quite separate and different to that of the building and functioning of a car. Something has to account for the interpolation of these two different classes of information into an interrelated and purposeful function. The anti-evolutionist will argue that this kind of association can only be arrived at by teleological influence; in the case of the garage, the person who built it. However, whilst the biology of life achieves the most masterful of engineering projects, they are not engineered as such. All such associations are arrived at on the basis of already existing natural, chemical affinities. The
160

gene has the capacity to recognise and incorporate potentially useful chemical affinities and to store them until such time as they become functionally associated by reason of enhancement of already ongoing systems.

On this basis, a very tiny little microbe evolved a very complex cyclical, chemical machinery, now known as the Krebs cycle, named after one of the German scientists who elucidated it in 1930. The Krebs cycle is a highly efficient method of producing ATP. It can produce some sixteen times the amount of ATP produced by fermentation per molecule of sugar. It is a kind of reverse autocatalysis and is the power unit now used by all aerobic life. This little microbe was engulfed or eaten by a larger microbe which failed to digest it. The little microbe found that it could live quite happily within the protective environment of the larger microbe and the larger microbe found the little microbe's propensity for producing ATP to be very useful. The two microbes then underwent a certain amount of gene transfer resulting in the whole becoming a single microbe with a membrane bound organelle, the Mitochondrion, inside it. However, the real magic of this communion lay in the fact that the mitochondrion retained its capacity to replicate independent of the larger microbe's replication machinery. This meant that the mitochondrion could go on replicating at a much faster rate than the combined organism as a whole, so producing a large number of mitochondria within the single celled microbe. This effectively flooded the single celled microbe with ATP power which, after some three thousand million years of life, finally shattered the restraining thermodynamic clamp. Life now had the power to expand beyond the single celled realm of being. All it needed now was the evolutionary potential. By use of the power now available, this was achieved in the following manner.

A particular microbe's DNA mutated in such a way that it interacted with its cell membrane chemistry and introduced a membranous material into the cytosol of the cell. It then further mutated, or more likely interpolated, to interact with this membranous material in a most astounding way. Its DNA molecule began to reorganise its gross configuration from a ring to a chromosomally sectioned, linear form and become encapsulated by the membrane to form the nucleus of the first eukaryotic cell. This represented a gigantic leap in the DNA molecule's capacity for information storage and handling. It was like jumping from a one hundred mega-byte CD to a ten terra-byte hard drive. It was from this potential, in conjunction with the new-found power to use it, that large multicellular organisms, including two-hundred-ton whales, were able to evolve.

Endosymbiosis, the engulfing of a small microbe by a larger one and the survival of both to their mutual benefit, is a practice in life which, in the way we have just seen, can dramatically affect the course of evolution. Another such event, where a microbe engulfed another microbe of cyanobacteria, initiated an evolutionary line establishing the growth of what are called plastids in the host cell. Plastids contain pigments compatible with the utilisation of solar energy and augment slightly different chemistries within the cell; changing its structure and manufacturing new substances like cellulose. This line of evolution eventually led to another split or divergence into the world of plants. In tidal areas and river estuaries, rudimentary plant forms began to creep onto the land. They colonised the land and diversified into many different species of plants able to utilise the most reliable energy source on Earth, the Sun. This radically changed the landscape of our planet and encouraged the development of amphibians. Some of these evolved into land bound animals with plentiful and easily obtainable food supplies. The way was now paved for the evolution of hundred-ton dinosaurs and seven- ton elephants, little shrews and, culminately, human beings with the capacity to contemplate the ruthlessness, the magic and the beauty of all that has gone before.

The foregoing, whilst being grossly abridged and corrupted by my own unqualified, personal leanings and interpretations, is, nevertheless, a reasonable layman's understanding of abiogenesis and its subsequent course of evolution. However, the philosophical interpretation of this truth on the part of science, at least as presented by one of its more eminent communicators, Professor Richard Dawkins, leaves a great deal to be desired.

The wrecking of the biblical stories by science together with its illustration of how our being is the result of chance encounters between chemical affinities, more complex but not greatly different, from that of a grain of sand, has led science to interpreting our being as one without meaning or purpose. This, together with its release from personal, disciplined responsibility is the root of growing atheism in our society. But science cannot deny that we are, by nature, communal and socialistic. So, there is something grossly wrong with this interpretation since clearly, it is a recipe for anarchy. To the modern atheist who would deny this, may I remind them that we are born with a God compartment in our minds and that they presently stand on a withering but deep cultural platform of values emanating from religious belief. Beware of its collapse into anarchy.

When I say there is a God compartment in our minds, this is not to associate it with the religious idea of God. But it is that part of our minds which elevates us from the compulsive behaviourism of the base animal. There is no set of man-made laws, any political system or regime, however democratic or oppressive that can equal, replace or imitate the God space in our minds. That space is an integral feature of the individual self and it is from there that we find our sense of purpose and generate our empathies and sense of moral values. I think it was Werner Heisenberg who said to Paul Dirac, "If there isn't a God, we should invent one." He was talking from that space. Atheism is a conscious shutting of that space and it is from there that Hitlers, Amins, Mugabes and other current oppressive and corrupt regimes emerge; the beginnings of our descent into anarchy. This is now also evident more colloquially. Fifty years ago, one could depend on our major institutions like banks, insurance companies, building societies and essentials providers etc. to be fair and just in their dealings. Nowadays, they make a serious and exacting science of extorting the maximum from their customer bases in return for the minimal possible service; the new world of so-called Financial Products. These are comprised mainly of mouse-trapping small print often used to submerge the commonest of sense. We have seen this attitude so severely applied in recent times that it has infringed on civil law and led to suits for compensation. This has also led to an even deeper greed and feed culture based on planned deception of bank status. Also, we have millionaires and billionaires whose wealth depends on human resource and its associated infrastructure yet they make a science of avoiding payment for it by tax evasion. There is no great difference in attitude nowadays between our higher institutions and organised criminality. A corrosive erosion of human dignity now follows the spread of atheism through human society. Atheism is now a fashionable choice and carries with it the most misconstrued idea that it displays some kind of expanded intellectual heroism. Nothing could be farther from the truth.

Chapter 8 - Religion and The Anthropic Principle

Firstly, I would ask the reader to specially note that whilst the word "Secular" or "Secularism" is used in the modern vernacular to describe people who are departed from religion and religious belief, this was not its original meaning. Its original meaning described people who objected to and departed from dogmatic religious authority. Most, if not all, of our early scientists were secular in this respect but retained their own personal religious leanings. I suspect that even Charles Darwin, in his abject mourning over the loss of his little Annie, consoled himself in the belief that such a person could not exist for nothing.

Irrespective of Professor Richard Dawkins, there is now a wind of similar leaning permeating the realm of modern science. It arises from the rapid improvement in the capacity of science to observe the workings and mechanistics of our wider universe and is described in what is called The Anthropic Principle. The reader should note that The Anthropic Principle derives from scientific interpretation and not from religion.

Anthropology is the scientific study of humanity, both evolutionary and socially. The Anthropic Principle is a scientific illustration of apparent purpose in our universe which is anthropologically oriented from its outset. The modern theologian has had to make an accommodation of science and welcomes The Anthropic Principle as a factual demonstration of our universe being a subject of omnipotent design. Notwithstanding this, the generality of science still opts for our universe being a subject of pure chance. This is not because of any enmity on the part of science towards the design concept, except perhaps for Richard Dawkins and his associates who have made a religion of New Atheism complete with the exact same baggage of sectarianism and fundamentalism as any other fundamentalist group. With the generality of the scientific community however, it is simply that the nature of science is such that it is bound by demonstrable, irrefrangible truth in so far as it is capable. At the same time, science would have to admit that the degree of chance it accepts is about as miraculous as the metaphysical idea and is also no more demonstrable. What we are left with then, is an impasse subject to philosophical interpretation. Any sane person must surely opt for purpose and meaning in their being rather than complete lack of it. That necessity is an innate feature of intelligence and many scientists satisfy it in the pursuance of their profession. The degree of merit in any such philosophising therefore, may be measured in terms of its conciliation of these extremes, its contribution to human social development in terms of purpose in our being and consequent departure from roads to anarchy. Science might also bear in mind that the Large Hadron Collider is far removed from the garden shed laboratory. Science is now entirely dependent on society.

Copernicus, followed by Galilei, clearly demonstrated, contrary to the general belief of the time, that the Earth did not occupy any central or privileged position in the universe. Together with Darwin's theory of evolution, it followed therefore, that neither did humanity. More modern and more highly sophisticated scientific revelations however; seem to indicate something quite different. It would seem that, whilst Earth indeed has no favour or centrality in terms of universal geography, it is highly favoured, even if by chance, in it's being the locus of what appears to be a central object of universal existence. This forms a natural leaning towards The Anthropic Principle.

Much of what immediately follows is difficult to explain in layman's language and is only understandable in terms of mathematics. The reader may therefore have to take a great deal for granted. I trust however, that I might manage to convey the essential point of the Anthropic Principle which is, that our universe seems to follow a logical process wherein its logic, including the mathematics to describe it, were innate qualities in the singularity from which it emerged.

The first thing to consider here is, what do we mean by chance? There are two kinds of chance. There is instituted chance and there is free or pure chance. The national lottery is a typical example of instituted chance. In other words, the lottery is a game that has been designed, its degree of chance calculated, its components organised and its activity consciously performed. A horse race is more or less the same. On the other hand, approximately one ten-centimetre cube of naturally evaporated sea water will eventually condense into a raindrop. Exactly where and when that raindrop will fall is a subject of pure or free chance. It is neither organised nor calculable. You can also have a mixture of instituted and free chance. Like, for example, if you go to a bookmaker during the month of July and place a bet on it's snowing in a particular area between six and ten o'clock on Christmas morning. The question is, from which, if any, of these three realms of chance did we and our universe emanate?

Big Bang theory is not the only but is the most widely accepted scientific theory of universal beginning. For the past half century, with increasing expertise in space technology and observational instrumentation, more and more empirical evidence has been piling up in its favour. It is an extrapolation of Edwin Hubble's discovery, in nineteen twenty-nine, that the universe, rather than being a steady state, is in a state of constant expansion and change. If this expansion is reversed, it leads back to a first instant of time when the expansion began, the Big Bang. There was no such thing as an instant of time before that event. Nor was there any such thing as a place or space where it could happen. Both time

167

and space are features of the event itself. Albert Einstein demonstrated that energy and matter are interchangeable and these concepts together lead to a scientific conclusion that, in the first instant of time, all that existed was an infinitesimal, dimensionless spot, or singularity, containing, in the form of pure energy in a state of extreme temperature, pressure and density, the entire universal potential. This singularity, suddenly appearing nowhere from nowhere, went into an immediate state of explosion. It was not an explosion in the sense that we see the random, destructive forces of a normal chemical explosion. It was an extremely rapid, light speed inflation of a pure energy state which, in itself, was highly specific and which expressed its specificity on both the nature and rate of expansion. This explosion could not happen randomly. It could only happen in the particular way in which it did, reflecting the specificity of the original singularity, features of which were space and time. This singularity was the only thing ever to have existed which was what the pre-Copernicans thought the stars and planets to be; a perfect Heavenly entity. It was pure, unused, unadulterated energy. Its expansion introduced the first entropic quantity.

In the first minute fractions of a second of universal expansion, what is called the Plank Time, or ten to the minus 43 second, (One trillionth of a trillionth of a trillionth of one ten millionth of one second) the singularity assumes its first sub-material state. Space and time initiated as a function of the prime energy state which expanded into them and expressed upon them a unified form of the four fundamental forces of nature. These are gravity, electromagnetism, the weak nuclear force and the strong nuclear force. During these first minute fractions of a second, the universe cooled from some immeasurable extreme to something in the order of a hundred million trillion trillion degrees C. Between ten to the minus 43 and ten to the minus 36 second, gravity separates from the other three fundamental forces and the most fundamental aspects of matter, the elementary fermions, which include the quarks and other leptons together with their antiparticles come into being. During the period ten to the minus 36 and ten to the minus

32 second, the strong and weak nuclear forces separate from the electromagnetic force and initiate a state of cosmological inflation. In these fleeting instants of time, the universe expands by an amount in the order of ten to the power of 28, an unimaginable, faster than light rate of expansion. This increases the size of the universe from something smaller than an atom to about that of a football and permeates it with a quark-gluon plasma. The rate of expansion now reverts back to something less than the speed of light. As the electromagnetic force separated from the other two, it created a state where, during the period ten to the minus 32 to ten to the minus 12 second, particle interactions were induced creating exotic particles including W bosons, Z bosons and Higgs bosons. The Higgs boson imparts mass on some of the particles, slowing them down. This allows the universe, hitherto comprised purely of radiation, to contain objects that have mass. By the time the universe reaches a trillionth of a second old, it has cooled to some ten quadrillion degrees C and has a density some one hundred million times that of lead. The four fundamental forces of nature have established their present form. Quarks, electrons and neutrinos together with their antiparticles are forming in huge numbers. Particles and antiparticles are annihilating one another but the process of Baryogenesis allows the survival of a sufficient number of normal matter quarks to form the present universe. (Baryogenesis is a hypothesis, based on quantum field theory, to explain the breaking of symmetry that allowed the quantity of normal matter to preponderate that of antimatter in the early universe) During the period ten to the minus 6 to the first full second of universal existence, the temperature drops to some one trillion degrees C. This is sufficiently cool to allow the quarks to combine in the formation of protons and neutrons, the hadron family of particles together with their antiparticles. These are the building blocks of the nuclei of conventional atoms. Neutrons are formed during this violent stage by the collisions of electrons with protons which also produces the massless neutrinos some of which recombine into new proton-electron pairs. This is an extremely

violent period during which, most of the hadrons and anti-hadrons annihilate one another; that is to say, they re-convert back from a material to a pure energy state. This may seem a complete melee but, in fact, it emerges as a form of control affecting the space-time - material - energy relationships of the overall Big Bang event. Electrons and their antiparticles, the positrons now comprise most of the mass of the universe. During the period one second to about three minutes, electrons and their antiparticles, the positrons, are colliding and annihilating one another and releasing energy in the form of photons, light particles. Colliding photons, in turn, recreate more electron-positron pairs. During the next fifteen minutes or so of expansion, the temperature of the universe falls to about a billion degrees C. This allows atomic nuclei of hydrogen and helium to form but there are still no atoms. The energies are still too high for electrons to combine with the nuclei. The universe is now about twenty-three minutes old and the temperature and density fall below that at which any more atomic nuclei can be formed. What was a dimensionless singularity has now expanded to universal proportion and is still expanding rapidly. From the beginning up until this time, particle energies have fallen in stages from some eight and a half million to two and a half million electron volts. At this stage, the universe is dominated by photons which are interacting intensely with charged protons, electrons and atomic nuclei.

Over the next quarter of a million to three hundred thousand years of expansion, the temperature of the universe falls to some three thousand degrees and particle energies to about two hundred and twenty thousand electron volts. This allows the nuclei of hydrogen and helium to capture electrons and so form the first gross material state, atoms of hydrogen and helium. Their relative abundance by mass was about seventy-five percent hydrogen to twenty-five percent helium. So, there were some twelve atoms of hydrogen to every atom of helium together with traces of lithium. With the electrons now bound up in atoms, the universe changes from being an opaque, soupy plasma to being transparent to light. It also releases the photons which, till now, have been reacting with electrons and protons in the photon-baryon plasma. The photons can now travel freely.

The next one hundred and fifty million years is sometimes referred to as the Dark Age of the universe. Although photons can travel freely at this time, there are, as yet, no concentrated photon emitters; there is nothing to shine. The universe is a fog of hydrogen and helium and is now relatively inactive with low energy and a vastness of time. It is still expanding and is being influenced by that strange and mysterious entity, dark matter.

At this stage, I believe it is of some benefit to attempt to convey to the lay reader some idea of the relationships and affinities existing between the apparent melee of particles created during the early instants of the Big Bang. These relationships derive from a precise, inbuilt specificity in the nature of the particles and determined the exact form and constitution of matter as we understand it. For the most part, these relationships are not explainable in layman's language and are only comprehensible in terms of advanced mathematics. But there is one set of particles which, whilst still having properties like colour, spin and others which are unexplainable in detail to the lay reader, also have a gross relationship which is understandable in layman's language. Hopefully, this will give the lay reader at least a glimmering of the degree of specificity, logic and beauty enjoyed by the physicist in his deeper understanding of the fundamentality of matter. Many scientists are moved more by this than art lovers are by a Picasso or a Rembrandt.

This class of particles is the quarks. There are six different kinds of quarks which science has named Up, Down, Strange, Charm, Bottom and Top. These names themselves are meaningless and only differentiate the six different kinds. The Up quarks and the Down quarks are of lower mass and the more massive Strange, Charm, Bottom and Top quarks, being unstable, tend to decay into the stable Up and Down quarks. Quarks do not exist as free entities in the evolved universe. They only existed as such during the extraordinary conditions of the early Big Bang. However, they are observable by scientists as free entities for fleeting instants of time in powerful particle accelerators like the Large Hadron Collider.

Up and Down quarks are highly specific in the respect that they are the building blocks or sub units of protons and neutrons. They interact by means of a force exchange boson called a gluon and combine to form the central mass of conventional atoms. The strong force, which binds atomic nuclear components together, is also an effect of the gluon force. What is called the elementary charge is the electrical charge carried by a proton. Throughout nature, with the single exception of quarks, it is always expressed in integer (Or whole number) values. In quarks, it is in fractional values of 1/3s and both neutrons and protons are each made up of three quarks. An up quark has a positive charge of +2/3 and a down quark has a negative charge of -1/3. A proton has the elementary charge of +1. It is comprised of two up quarks and one down quark. This gives a proton the composite fractional charges of:

(+2/3) + (+2/3) + (-1/3)

This resolves to 2/3 + 2/3 equals 4/3 minus 1/3 equals 3/3 or +1, the elementary charge. The neutron is comprised of two down quarks and one up quark. This gives it composite fractional charges of: (-1/3) + (-1/3) + (+2/3) This resolves to -1/3 and -1/3 equals -2/3 which is cancelled out by +2/3 equalling 0 or zero electric charge.

Whilst quarks and these simple arithmetic values relating to them may be of no significant interest to the lay reader, the essential point is, they do not occur randomly. They are highly specific both in their form and physical interaction. If any one of them had been different, neither matter nor you nor I would exist.

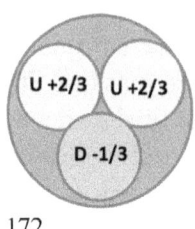

So, the Big Bang was not a melee. It was a clear pursuance of logical process rather than a chaotic state. The particular rate of

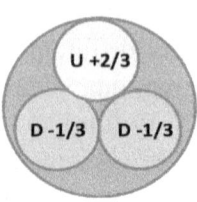

expansion caused a graduating change in energy volume, density, pressure and temperature. This, in turn, caused a specific range of changes in energy states which was expressed as a co-transience of energy to matter in the form of the fundamental quantum particles.

During the dark age of the universe, some anomalies in homogeneity generated during the brief inflationary period express themselves as an inhomogeneity in the vast, expanding gas volume. (There is still some debate within the scientific community regarding inflation and its effects. However, the universe is as it is. Rather than the randomness arising from chemical explosions and the like, the universe is highly ordered. Inflationary theory was devised to account for this and arrives at what is called the Copernican or Cosmological Principle. This states that the universe is both Isotropic and Homogeneous. Isotropic means the universe is the same wherever it is viewed and from wherever it is viewed; there are no special directions to the universe. Homogeneous means the universe is the same in all localities; there are no special places in the universe. However, this is a gross or largest scale view. A more fractal view reveals that this gross uniformity comprises a universe of local differences. It is this ingenious combination of fractal within a general uniformity that allows for individual galaxies of stars and for individual stars within galaxies). If everything was exactly smooth and even then, all forces would be in equilibrium and nothing would happen.

Over the next one billion years, the universe becomes subject to a tug-of-war between expansion and the force of gravity acting on these anomalies. This causes the unimaginably huge volume of relatively dense hydrogen and helium to fragment into huge and separated clumps which, due to continuing universal expansion and some as yet undetermined relationship between gravity and dark energy, will determine the material distribution and future galactic geography of the universe. In the early universe, gravity acted on these huge clumps of gas so as to cause an in-fall towards their centres of such immense concentration that black holes were created. These black holes were so massive, in the order of billions of solar masses, that interactions between the components of the in-falling material generated what are called Quasars. Quasars are amongst the most powerful objects in the universe. They can consume an amount of material equivalent to that of the entire Earth every minute. They radiate over the entire electromagnetic spectrum with sufficient intensity to slow down the rate of in-fall. Also, the immense gravitational influence of the black hole initiates a rotational movement in the overall gas cloud, not quite so fierce but something like the way your bath water rotates as it goes down the plug hole. This causes currents and eddies to develop within the cloud which, in turn, causes volumes of different gas densities to develop. These have their own gravitational centres which attract more material and initiate the formation of a galaxy of stars. They were the first shining, causing our universe to come alight. All galaxies, including our own, appear to have a super massive black hole at their centres but not all of them have quasars. Quasars seem to be more a feature of the younger, denser universe although, there are some known to be only three billion years old. There are less than a quarter of a million quasars known to science in a universe of some two hundred billion galaxies each with an average of two to four hundred billion stars. In the younger universe, which was dense, stars tended to be very massive. This meant they had a huge amount of gravitational energy which promoted enormous nuclear fusion power; sufficient to create elements heavier than helium and up to iron. The power and speed of their fusion rates caused them to have very short lifetimes. They died by means of supernovae or gamma ray explosion during which, even heavier elements were created. The supernovae explosion could distribute these elements over galaxy sized regions, gravitationally and materially seeding them for birth of new stars; a kind of begetting of stars by stars.

In the foregoing, The Anthropic Principle sees the series of events emanating from the Big Bang as a logically interrelated series where each stage was preparatory to each ensuing stage; much like the building of any machine or process. It culminated in a gross but simple form of matter i.e. hydrogen and helium. The highly complex chemistry of life would be impossible by any manipulation of these relatively simple elements. So, hydrogen and helium were not an end but merely mark a particular and important stage in what is still an ongoing process. The hydrogen atom is the basic building unit of the wider range of heavier elements and stars, fuelled by gravity, are the incredible machines that perform the transmutations by a process called thermo-nuclear synthesis.

Chapter 9 - Mechanisms of The Stars

Thermonuclear synthesis was theorised by the British astronomer and mathematician Sir Fred Hoyle in the late nineteen- forties. It described how the whole range of elements was manufactured in stars using the hydrogen atom as the basic building block. At the beginning of the nineteen-fifties, he also predicted the phenomenon of a particular resonance energy matching which facilitated the otherwise problematic making of carbon in stars; the Triple Alpha Process. He then worked throughout the fifties with the brilliant American physicist, Professor William Alfred Fowler, in consolidating and proving the theory.

Stars are categorised into three groups by age and called populations 1, 11 and 111. Population 111 stars are a hypothetical group since none have yet been observed. They are believed to

Sir Fred Hoyle and Professor W.A. Fowler

have been the first stars to have appeared in our universe. As such, they would have been very massive and would have had very short lifetimes. We shall see later however that even these played an essential part in a closely related and intricate universal mechanism. Population 11 includes the oldest observable stars and population 1 comprises the young and what are called "Main Sequence Stars. Our Sun is a main sequence star.

As described earlier, the huge gas clouds in which stars are born and develop are not uniform either in volume or density. They are constantly subject to influences of other stars, including shockwaves, which cause various currents and eddies to form in the clouds. These are areas that are effectively being pressurised and so form their own various gravitational centres which results in different types of new stars being formed. The fundamental driving force of star machinery is gravity. Gravity is proximity and mass related so a star type is determined by the volume and density of the local gas medium in which it forms. In other words, the more material available within its gravitational influence during the birth and growth of a star, the more massive the star will become and the greater its gravitational energy will be. Where gas clouds are volumous and dense, numbers of stars can compete with one another in procuring material. This again, may seem a disordered melee of events. However, we shall see that a range of different star types is necessary in order to produce life. If all stars were of a single type, it is doubtful life would have been possible. The Anthropic Principle sees this as a continuance of a teleological systemised process starting at the Big Bang and directed towards the creation of life and culminately, human intelligence. This however, is a theological interpretation. We shall see later that there is a subtler, if somewhat disappointing, but more realistic and rational interpretation; one quite devoid of magic and miracle.

Scientists have classified the different star types in accordance with what is called their "Spectral Characteristic" or spectrum of the light they emit. The normal daylight we see is actually a mixture of colours (Or frequencies) of light. If you shine it through a prism, as Isaac Newton did, the prism separates the various frequencies and you can see the different colours. In certain weather conditions, the moisture in the air can act like a prism and produce a rainbow. A rainbow is the spectrum produced by the Sun which is a common type of star.

Astronomers classify the stars by colour and designate them by the letters O, B, A, F, G, K and M. O stars are the hottest and M stars the coolest. The difference between one class and another is not a quantum jump. There is a graduation of difference between the classes. This is stated as a number in the classification representing tens of percent of the difference. The Sun is classified as a G2 star. This means it is a G star which is two tens or twenty percent along the way to being a K star. A roman numeral is used at the end of the classification to describe the size and luminosity of the star. They range from 1 for super-giants to V for dwarfs or main sequence stars. Thus, the Sun is classified as a G2V star.

HOTTEST				Sun		COOLEST
O	B	A	F	G	K	M
BLUE						RED

The difference in classes of stars arises as a consequence of their different masses and resultant gravitational pressures which result in different core temperatures and luminosities. The unit of measurement for star mass and luminosity is the mass and luminosity of our home star, the Sun. Hence, the masses and luminosities of other stars are stated in terms of the mass and luminosity of the Sun, the Sun being 1 solar mass and 1 luminosity.

To give your mind some idea of what this is, consider the Earth with a mass of some one thousand trillion metric tonnes. Our largest planet Jupiter is three hundred and eighteen times the mass of the Earth and the Sun is one thousand and forty-eight times the mass of Jupiter or a third of a million times the mass of the Earth.

One of the smallest stars known to astronomers is the secondary in a binary system known as Gliese 623b. It orbits its primary partner once in 3.7 Earth years and has a mass of 0.11 or just over ten percent solar mass. This is near the verge of the minimal possible mass for a star. A little smaller, and it would have had insufficient gravitational pressure to ignite as a star. It would have been a brown dwarf. As it is, it is a red dwarf, one of what is now believed to be the most numerous and long-lived stars in the universe. The largest known stars, of which there are very few, are in the order of one hundred or more times the mass of the Sun. The Sun is considered to be a medium sized star.

Atoms and the ways in which they combine by means of molecular cohesion, ionic and covalent bonds is what accounts for the widely varying range of substances we see around us and we are pretty well capable of distinguishing between different substances. You don't need to be a chemist to know the difference between metal and non-metal substances. Astronomers however, use the word "Metal" in a completely different sense. Chemistry does not occur in stars and astronomers use the word metal to describe any element in the atomic table above helium. So, in astronomy, you have hydrogen and helium as the two lightest (Or lowest mass) atoms and all the rest are metals even though they might be gasses.

Different star types produce different and different amounts of metals and this is referred to as their "Metallicity" Metallicity is one of the more important properties of stars in terms of life potential. It comprises the substances from which rocky planets and complex molecules can form. Metallicity may, at first sight, seem a little confusing to the lay reader. This is because the massive stars that produce most of the metallicity are referred to by astronomers as low metallicity stars whilst medium stars like the Sun, which produce little or no metallicity, are referred to as high metallicity stars. The reason for this is the stars that produce metallicity are generally massive and shorter-lived stars. The amount of metallicity they produce is very small in comparison to their overall mass. Also, because many of them were older stars, the interstellar medium in which they formed contained little or no metallicity. However, these massive stars were shorter lived so many of them were born, lived and died over billions of years before stars like our Sun were born. As they died, they contributed their metallicity to the interstellar medium so, over these billions of years the interstellar medium itself became richer and richer in metallicity. Our Sun therefore, was born and formed from a medium which was already relatively rich in metallicity. This is why, even though it produces very little metallicity itself, our Sun is a high metallicity star. Its metallicity represents some 1.6 percent of its mass

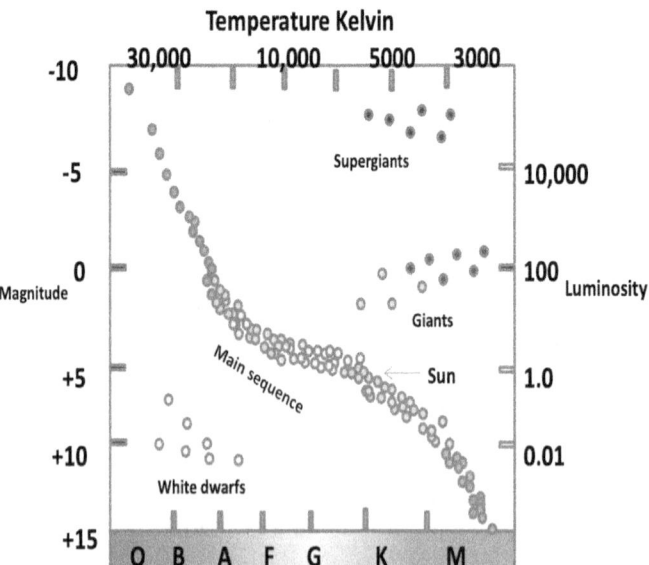

Fig. 24 Hertzsprung – Russell diagram

All stars are born in more or less the same way but their lives differ in accordance with differences in the volumes and densities of the local gas clouds from which they form. Fig. 24 shows the Hertzsprung-Russell diagram, named after the scientists who constructed it and normally referred to as the HR diagram. It shows the main parameters of the stars and the birth, life and death of a star can be followed on this diagram. It is one of the most widely used tools of astronomers and astrophysicists. The lay reader should note that the smaller the magnitude, in the left-hand column, the brighter the star. Since magnitude and luminosity are both measures of light, this may seem a little confusing. The reason for this is, the luminosity is the calculated amount of energy actually being radiated by a star as opposed to magnitude which is how the star appears to the eye. A dimmer star that is close may look brighter than a brighter star that is far away.

The huge clouds of gas from which stars are formed are called giant molecular clouds. This is because the hydrogen of which they are mainly composed does not comprise single atoms but atoms that are bound into H_2, two hydrogen atoms bound together into a molecule. This provides both atoms with their happy state of having two electrons in their single orbital. The clouds are very large. It would take light, travelling at three hundred thousand kilometres per second, some tens to hundreds of thousands of years to cross a typical giant stellar molecular cloud (The distance light travels in one year is called "A light year") and that cloud could contain sufficient mass to make several galaxies containing several hundred billion stars the size of our sun. You will already be aware that when you heat a gas, it expands. Conversely, when you cool it, it contracts in volume and becomes denser. The temperature of interstellar molecular clouds is about ten degrees above absolute zero. This is very cold indeed so the clouds tend to contract to a density determined by their very low kinetic energy and, in doing so tend to develop a rotational movement. By this means, the cloud assumes a maximum natural density throughout. That is the maximum density it can achieve without being pressurised. Influences of the cloud's rotational movement and of stars either within or near the cloud in the form of shock waves or gravitational

181

disturbance cause currents and eddies to develop within the cloud. These currents and eddies are effectively areas that are being pressurised so they contract further beyond their natural density. This means that more gas is being compressed into a smaller volume so its mass becomes greater than the gas in the immediately surrounding area. This establishes a gravitational centre which exerts a gravitational pull on the surrounding gas, pulling more and more gas towards its centre. It also speeds up the rotational movement of the cloud in its local area. This causes the cloud to fragment into huge clumps of varying sizes measuring from tens to many hundreds of light years across. It is these clumps that will form into galaxies of stars. These galactic sized clumps of gas undergo a similar process and fragment further into the smaller clumps within which stars are born.

So how does one of these clumps form into a Sun like star? This particular volume of gas, having established a gravitational centre, begins to contract under its gravitational influence. This forms a core of a young, embryonic star. The core starts to grow and, by growing, becomes more and more dense and more and more massive. The more massive it becomes, the greater is its gravitational influence, not only on the surrounding gas, but also in further pressurising the gas already in its core. This increasing pressure at its core brings the atoms of the gas into closer and closer proximity causing the atoms to collide more and more frequently and with ever increasing velocities. This is expressed as an ever-increasing temperature at the core. This whole process continues as long as there is a sufficient amount of gas and dust available within its gravitational influence. However, from this stage in its growth, the dynamics of its further growth become rather complex and are, as yet, not fully understood. The embryonic star's increasing gravitational influence causes it, as well as the part of the molecular cloud from which it is forming, to start spinning. This converts some of its gravitational energy into angular momentum. Angular momentum is a conserved property so the infalling material adds to the young star's angular momentum, causing it to spin faster. This generates centrifugal force, expressed mainly around the equator of the young star. As a result, the star's globular or spherical configuration begins to change. It begins to flatten out along the plane of its equator into more of a disc shape than a sphere. It still retains a globular centre and continues to accrete material both from the surrounding medium and from the inner edges of the disc. The central sphere is still growing. It is becoming more massive and getting hotter and hotter. The core of the star is now at a temperature of some three to four million degrees. It is still not hot enough to fuse hydrogen into helium but it is sufficiently hot to force protons into the nuclei of lithium atoms destroying the lithium and producing four helium nuclei. The star is now in its T-Tauri or pre-main sequence stage. It first becomes visible as a brown then a red dwarf due to the energy being produced by gravitational pressure. It is still growing and getting hotter. As its core temperature reaches between two and a half and four million degrees, it starts to burn lithium and the destruction of lithium releases a great deal of energy causing the young star to become very bright. It is spinning quite rapidly, completing a revolution in one to ten or so Earth days. (Our Sun spins only once in a month). It dissipates its momentum energy by energetic expulsion of material from its poles. This appears as energetic columns shooting out from both poles at right angles to the star's plane of rotation and to the disc. Eventually, when this growing body begins to approach about ten percent the mass of our Sun, the core temperature reaches some ten million degrees C. At this temperature and pressure, hydrogen atoms are charging around at incredible speeds.

It needs to be explained here that, whilst we are talking about atoms, the atoms in the core of a star are not really complete atoms. At star core temperatures, the atom's electrons have been stripped away so that the atoms we are talking about here are actually positively charged ions. You are already aware that like charges repel one another. So, in the core of a star, this repulsion represents a force preventing the ions from actually colliding. This is called the Coulomb force. At temperatures in the region of ten million degrees, the incredible velocities of the hydrogen atoms are sufficient for them to punch through the electrostatic repulsion of the Coulomb force by a means called quantum tunnelling. (Unfortunately, any explanation of quantum tunnelling is beyond the scope of this work and is not necessary in terms of the object of the work. It can only be explained mathematically and in conjunction with an understanding of quantum mechanics. Suffice to say here that it is to do with the wave-particle duality of subatomic particles and their capacity, when encountering an energy potential barrier, to absorb sufficient energy from the barrier to enable them to pass through it rather than having to climb over it.) The nuclei of the hydrogen atoms, having quantum tunnelled through the coulomb force, are now forced into the fields of one another's Strong Force (The nuclear binding force) which is attractive and fused together to form helium. This process releases a tremendous amount of energy and the star ignites as a full-blown star. It enters the HR diagram where shown by the dotted line in Fig. 25. You can see from this that, at this stage, the Sun, now being about a hundred million years old, is more than a hundred times brighter than it is now. This is because it was much larger in radius at that time.

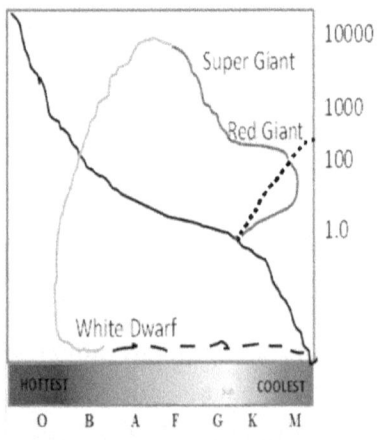

Fig. 25 Track of Sun life cycle

As the Sun traverses the dotted line from right to left across the HR diagram, it is still accreting material both from its surrounding disc and from the molecular cloud. It is becoming more massive and its gravitational pressure is increasing. Its core is becoming hotter and its outward radiation pressure greater. Also, protoplanets and possibly even organic molecules are now forming from the metallicity within the disc. As the Sun's mass increases, its gravitational pressure increases. This causes greater and greater compaction of the star and begins to reduce its size. Its radius is now getting smaller. Its core temperature is climbing to over ten million degrees. Its outward radiation pressure is increasing and this is producing an ever- increasing powerful solar wind which is beginning to dissipate its surrounding disc and remains of the stellar cloud from which it formed. At this stage, the Sun is maturing from a T-Tauri to a main sequence star. It reaches a stage where its outward radiation pressure exactly balances its inward gravitational pressure and it settles on the main sequence in a state of what is called hydrostatic equilibrium. It will not grow any further. It will remain on the main sequence where it will transmute some six hundred and fourteen million tons of hydrogen into helium every second for the next ten to fifteen billion years. The stability of the Sun on the main sequence is almost exactly like a man-made thermostat. If the rate of fusion in the core reduces, the core will cool and reduce the outward radiation pressure. This will allow gravity to further pressurise the core so increasing the rate of fusion and producing greater radiation pressure and so maintaining the state of hydrostatic equilibrium.

The black changing to grey line in Fig. 25 shows the progression of the Sun through the HR diagram as it moves off the main sequence and enters its dying years.

As already mentioned, in a star like the Sun, with a core temperature in the region of fifteen million degrees, the electron normally moving around the nucleus of the hydrogen atom is stripped away so that the hydrogen atoms are ionised and positively charged. They therefore repel one another by the Coulomb force. But, at this temperature, the nuclei are charging around at such incredible speeds that, when they collide, they quantum tunnel through the field of the repulsive Coulomb force and into the very short-range nuclear binding force, the Strong Force. The Strong Force is the most powerful of the natural forces and is an attractive force so it fuses these nuclei together but not in a random fashion. The result of these random collisions is a specific arrangement which reflects the specificity (Or characteristics) of the particles and forces involved. Nuclear fusion in a star like the Sun follows a procedure described by physicists as the proton proton or pp chain reaction. In this reaction, four atoms of hydrogen follow a procedure which fuses them together to form one atom of helium. Four atoms of hydrogen have a slightly greater combined mass than one atom of helium so, during this reaction some mass has actually gone missing. This missing mass has been re-converted back to energy in accordance with Einstein's famous equation $E = Mc^2$. In this equation, E is the energy produced, M is the amount of mass converted and c squared is the amount by which the mass is multiplied to obtain the value of E. In the hydrogen pp reaction, the amount of mass converted is very small but the value of C squared is ninety billion so the amount of energy produced is considerable; it is about the same as that generated by burning an average size shoe box. Also, this reaction involves only four hydrogen atoms but the Sun is converting six hundred and fourteen million tons of hydrogen into six hundred and ten million tons of helium every second. This amounts to billions of billions of quadrillions of atoms and, as you can see, four million tons of matter is converted back into energy every second. This is why a star like the Sun can radiate such an incredible amount of energy in the form of light and heat.

Just to give you some idea of what this is, take a white standard size football and put a single black spot from the tip of a felt pen on its surface. Now look at the whole. The surface of the football represents the radiation sphere of the Sun and the black

186

spot represents the miniscule amount of it captured by the Earth. Even this relatively tiny amount is more than ten times what we would ever need if we could but utilise it. The Sun is an incredibly powerful machine.

Obviously, scientists cannot observe the life of a star in the way just described. The period of time over which imaginatively thinking humanity has existed, some forty thousand years, is millions of times less than the winking of an eye in the lifetime of a star. However, light has a finite speed of three hundred thousand kilometres per second and interstellar distances are vast. When scientists are examining a star, they can look at it by means of radio telescopes. This means they can look at it in a number of different frequencies as well as visible light. By this means they can determine, quite exactly, different star types and even much about their composition. They can look at two stars a million light years apart and tell if they are of the same type. By calculation involving the speed of light they can tell the distance between them and if one is a million light years farther away than the other then they know they are looking at it a million years earlier in time. They are seeing it at an earlier stage in its life. By this means they can observe a large number of stars at various stages in their lives and can accurately construct the foregoing scenario from these observations.

The dying years of a Sun-like star are complex and require some knowledge of the Ideal Gas Law together with some mathematical understanding. This is in the form of a mathematical operator known as virial theorem and in conjunction with a rather non-intuitive phenomenon known as negative heat capacity. When you inject heat into a system, it normally gets hotter. However, there are circumstances where the injection of heat causes a rate of expansion which causes the system to cool. This is expressed as a negative heat capacity of the system.

Even after ten or fifteen billion years, a star like the Sun has only lost a very small percentage of its original mass when it joined the main sequence. It now enters a phase where, thermodynamically speaking, it changes from a single fully integrated system into two sub-systems, the core and the gaseous envelope, each of which follow completely opposite thermal directions. The core contracts and gets hotter and the envelope expands and gets cooler. The Sun is now turning into a red giant. It will become so large that it will engulf the inner planets Mercury and Venus and will even approach the Earth to a distance where its three-thousand-degree surface temperature will fry the Earth to a cinder.

By this time, most of the Sun's hydrogen fuel has been used up and its rate of hydrogen fusion begins to decrease. Also, hydrogen is lighter than helium so stratification of the core has taken place with helium sinking into the core and the remaining hydrogen forming a shell around it. The star's outward radiation pressure begins to decrease so its tremendous gravitational pressure begins to further pressurise the helium core as well as the remaining hydrogen around it. The core becomes so compacted that its gravity now encounters another resistive force. This is called Electron Degeneracy. Electron Degeneracy is a force described by the Pauli Exclusion Principle which states that no two fermions within a system can occupy the same energy states. At this stage, the Sun's core is so compacted that all the available electron energy states are occupied. The core cannot be compacted any further and it cannot expand so it just gets hotter and hotter. Degeneracy is the densest state of matter achievable by a Sun-like star. An egg cupful of helium degenerate matter would weigh more than a London bus and a soup bowl full as much as a large steam locomotive. Injection of heat into degenerate matter does not cause it to expand. It just gets hotter and hotter. The temperature of the core now increases to over fifteen million degrees.

At this temperature, another nuclear reaction is initiated. You will remember that, during the initial formation of the Sun, it was endowed with a small amount of metallicity from the gas cloud in which it formed. Included in this were certain abundances of carbon, nitrogen and oxygen as well as other metals. At over fifteen

188

million degrees, the carbon atom acts as a catalyst and initiates a cyclical, catalytic, nuclear reaction called the CNO or carbon-nitrogen-oxygen reaction. This is a complex, looped ring of reactions catalysed at different stages by carbon, nitrogen and oxygen and during which, four atoms of hydrogen are transmuted into one atom of helium. The catalysing agents also still remain.

This process is relatively short lived in the overall life of the Sun. But it generates a lot of heat which heats the core even further and causes the envelope to expand and cool into an even larger red giant. Since the core cannot expand and can only get hotter, it eventually passes seventeen million degrees where the CNO reaction becomes much faster. The core is heated even further and begins to approach a hundred million degrees. This initiates another kind of reaction called The Triple Alpha Process. It transmutes three helium nuclei into one nucleus of carbon 12. As a side effect, some of the new carbon nuclei will absorb another helium nucleus and become oxygen 16. In the carbon 12 and oxygen 16 here, the numbers 12 and 16 do not concern us. They are what are called the mass numbers of these particular atoms.

At over one hundred million degrees, the Triple Alpha Process is the beginning of helium burning and transmutes helium into carbon and oxygen; the most essential ingredients for life. It is a very powerful reaction. At the same time, it is a highly improbable one; so improbable that the Anthropic Principle sees its happening as a kind of tweaking by the hand of God. But Sir Fred Hoyle was not God; just a brilliant mind dedicated to science.

So how does a highly improbable nuclear reaction, one that is unlikely ever to happen, become a feature of star machinery and produce the fundamental basis of life i.e. carbon? In order to understand this, you would need some understanding of physics but you may think about it in the following manner.

When we say that nuclear fusion releases energy, this energy is radiated, depending on the particular reaction, in the form of neutrinos or alpha or beta particles or gamma rays or other forms of electromagnetic radiation. (A helium nucleus and an alpha particle

189

are one and the same thing) In other words, the radiation is not random. It comes in specific amounts called quantum.

At this stage, I would refer the reader back to where we were considering the formation of molecules from atoms. You will remember that this came about by ionic and covalent bonding. You will also remember that ionic and covalent bonding came about because of atoms having happy and unhappy states and their tendency to be happy constituted the basis of the science of chemistry.

The nuclei of atoms also have happy and unhappy states. But, unlike the states of whole atoms, any explanation of the happy states of nuclei is highly technical and beyond the level of our discussion here. It involves things like magnetic moment, spin numbers, shells with what physicists call magic numbers and so on. Suffice to say here that the nucleus of an atom is happy when all of the persuasions emanating from the compatibilities of its interacting component parts and forces are satisfied or balanced. Scientists call this the Ground state or Natural Resonant Energy of the nucleus. Each different kind of atom has its own resonant energy. If the energy of an atom is greater than this, it is said to be in an excited state and will undergo what scientists call Decay. Decay is when an atom radiates energy in order to arrive at its ground state or natural resonant energy.

The reason why the triple alpha process needs such high temperatures to initiate is because the helium nucleus has two protons as opposed to hydrogen which has only one. The coulomb force of helium is therefore very much greater than that of hydrogen. Nuclei of helium therefore need to be travelling at much higher velocities than hydrogen nuclei in order to get within quantum tunnelling range. Also, if the fusion of nuclei is to occur, the nuclei have to collide head on in order to utilise the whole of their kinetic energy; otherwise they will simply bounce off one another.

However, like the pp chain and CNO reactions, the triple alpha process occurs in steps or stages and quite slowly at first. In the
190

first step, two helium nuclei fuse together to form an isotope of beryllium, beryllium-8. This isotope is very excited and decays back into helium by radiating two alpha particles (Or helium nuclei) in just ten to the minus fourteen or fourteen trillionths of a second. The chances of a third helium nucleus being absorbed within that fleeting amount of time are highly improbable indeed. However, it just so happens that the sum of the resonant energies of one helium nucleus and one beryllium- 8 isotope is exactly equal to that of a mildly excited state of carbon. This makes the beryllium isotope highly receptive to any incoming helium nucleus because it is just as happy to receive the helium nucleus and become stable carbon as it is to decay. The angle of collision does not need to be anything like so accurate. This changes the chances of a successful collision from one of very high improbability to one of reasonable probability and the excited carbon atom emits a gamma-ray photon, decaying into stable carbon 12. (A similar kind of resonance relationship between carbon and magnesium facilitates the burning of carbon in more massive stars). Even so, the rate of carbon production is very slow because the beryllium isotope decays so quickly; there just aren't enough of them around. However, some carbon is being produced and the heat from this together with that still being produced by the outer shell of hydrogen fusing into helium is further heating the degenerate core. The triple alpha process is very highly temperature dependent. The higher the temperature, the faster the atoms are moving and the faster it goes. If the core temperature were to be doubled at this stage, the triple alpha process would run a trillion times faster. The core temperature doesn't double but it does eventually get high enough for carbon to be produced at a faster rate than the rate of decay of the beryllium 8 isotope. This is a dramatic turning point in the whole process.

It is now over a billion years since the star left its main sequence. As we have already seen, when dealing with stars, we are generally dealing with events that take from millions to billions of years. In the dying years of a Sun-like star, when the triple alpha process reaches the stage where carbon is being produced at a faster rate than beryllium 8 is decaying, it suddenly becomes a runaway reaction. Some sixty to eighty percent of the star's entire helium content is transmuted into carbon over a period of only two minutes. This amounts to a burning rate of about ten Earth masses per second and is an unimaginably gigantic explosion. It produces as much light as a whole galaxy of stars and as much energy as a Sun-like star produces over two hundred million years on the main sequence. It is called "The helium flash" and is described by many scientists as the final bullet in the heart of a dying Sun. The helium flash does not create a spectacular sight because it is so deeply buried and obscured within the gigantic red giant envelope.

In fact, by human standards, a Sun-like star at this stage still has a bit of life left in it; albeit late retirement rather than useful in terms of its primary function.   When its core first became degenerate, the automatic thermostatic control affected by hydrostatic equilibrium was switched off. The core was compacted into a white dwarf and the envelope expanded to red giant proportions. But the incredible explosive power of the helium flash is sufficient to lift the core out of its degenerate state and back to an incredibly dense, expanded and cooling gaseous plasma. In order to do this, the helium flash explosion has lifted the equivalent of a hundred thousand earth masses of degenerate matter up to several times its white dwarf volume. Little or none of the energy of the helium flash reaches the surface of the red giant. All of it is consumed in this tremendous weight lifting feat.

The effect of this is a rapid cooling of the expanding core which also cools the remaining helium and the hydrogen surrounding it. This in turn reduces their fusion rates and dramatically reduces their outward radiation pressures. Gravity again kicks in and re-couples the core and the gaseous envelope into a unified thermal system. The thermostatic effect of hydrostatic equilibrium switches back on. Over a short period of time (For a star that is), about ten thousand years, the red giant shrinks by about ninety-eight percent of its volume. The whole changes from a white dwarf with a departing envelope into an integrated sub-giant, yellow orange star about ten times the size of the present Sun and forty times as bright.

The star will now settle for a time on what is called the Helium main sequence on the HR diagram. However, helium burning produces less than ten percent the energy of hydrogen burning so the star is still essentially a hydrogen burning star. But it now has a carbon core surrounded by a helium shell which in turn is surrounded by the remaining hydrogen shell which is still fusing into helium and helium in turn into carbon.

In order to maintain its renewed state of hydrostatic equilibrium, the star now has to burn its remaining hydrogen about a hundred times faster than when it was on the main sequence. This soon depletes the already starving hydrogen shell and reduces the outward radiation. This, in turn, allows gravity to again compact the core which again starts to become degenerate and increase in temperature. The core and the envelope again thermally decouple. The core gets smaller and hotter and the gaseous envelope gets larger and cooler. It is now ascending again to red giant proportions. The core becomes hot enough to initiate another helium flash but this time, of much less power. This process will now continue for about half a million years with successively smaller helium flashes occurring about a hundred thousand years apart and blowing the outer gas envelope away from the core altogether. The envelope will now continue to expand and cool and become more and more diffuse as it moves away into the interstellar medium in the form of what eighteenth century astronomers called a Planetary Nebula. Meanwhile, the incredibly dense and degenerate white dwarf carbon core has reached a brilliance of some four thousand solar luminosities which lights up the departing nebula and provides one of the most spectacularly beautiful sights in all of nature. It is like a galactic size fireworks display celebrating the life of the star; one which was moderate in temper and provided the goldilocks conditions for long enough (In the case of our Sun) for a simple chemistry to evolve into highly complex and intelligent life.

Helix Nebula: Remnant of a Sun-like Star.

The white dwarf core will now move down the left-hand side of the HR diagram as it slowly cools over many tens of billions of years. The reason it takes so long to cool is because sixty percent of a solar star mass has now been reduced to a surface area of Earth size proportions. It will eventually follow the black dashed line along the bottom from left to right and move out of the HR diagram as a black, invisible carbon cinder. This has not yet happened to any Sun-like star because it takes so long that the universe is not yet old enough. The oldest of these dwarfs are still glowing in the bottom left corner of the HR diagram. In the event that the dwarf, white or black, should ever come into close proximity or collision with another star, it can re-activate and merge with its new companion. In that case, it will live again for a time and die again but in a different way together with its companion. In such a case, it may spread its metallicity back into the interstellar medium. Otherwise, its metallicity will remain locked forever in the grip of an incredibly powerful gravitational vice. Such mergings do occur especially if the dwarf was a companion in a binary or multiple star system where it can become involved in a type 1a supernova explosion.

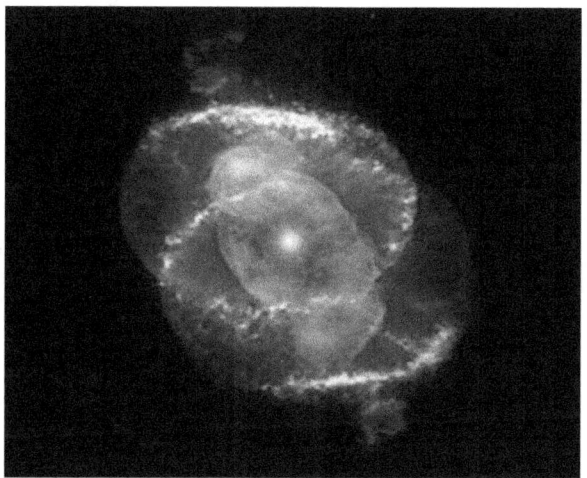

Courtesy Wikimedia commons
Cat's eye nebula: Remnant of a sun-like star

What we have just considered, whilst again being grossly abridged and not entirely accurate in all respects, is nevertheless a reasonable layperson's understanding of the principles of star machinery. All of the different types of stars work in more or less the same way but their lives, deaths and what they leave behind differ in accordance with their different masses.

We have seen how a star's mass depends on the size and density of the molecular cloud from which it forms. We have seen how a Sun-like star transmutes hydrogen into helium which then sinks into and forms a helium core. We have seen how, as the core gets hotter, it then transmutes helium into carbon which, in turn, sinks into and forms a carbon core. This becomes degenerate and gets hot enough to initiate the helium flash which blows away the envelope, leaving behind a white dwarf about the size of the Earth.

The foregoing scenario is typical of stars from about a half to six times the mass of the Sun. Stars more massive than this have sufficient gravitational energy to transmute carbon into neon, neon into oxygen, oxygen into silicon and silicon into iron. These build up in successive shells just as the carbon, helium and hydrogen did in a Sun-like star. Scientists refer to these as onion layered stars; see Fig. 26. The most massive substance, iron, forms the core with successively lighter elements forming shells around it. Iron marks a very particular line in the transmutation process. This is because it is the most stable and therefore most tightly bound of all the elements. As such, the transmutation of iron would be an endothermic rather than an exothermic reaction. In other words, the transmutation of iron does not produce energy in the star. It actually uses up the star's energy. These more massive stars do not produce a helium flash. This is because their greater gravitational energies produce the necessary temperatures for higher element transmutations before the core becomes degenerate. Carbon burning requires temperatures in excess of a billion degrees.

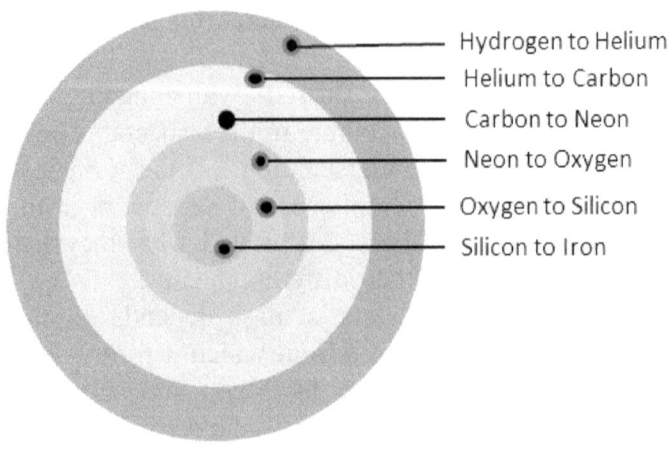

Onion layered star

Fig. 26 Onion layer arrangement of high mass stars.

Stars of six to ten solar masses have unimaginably powerful gravitational energies and can have core temperatures in the order of three billion degrees. As a result of this, they burn their hydrogen fuel at a faster rate than lesser stars. The higher elements above helium require successively higher burning temperatures, their burning periods are successively shorter and their contribution to the star's outward radiation pressure successively less. Also, huge and powerful convection currents develop within these stars. These currents dredge up and distribute near core elements into the general body of the envelope. A stage is reached where their hydrostatic equilibrium is lost and gravity completely overpowers any radiation pressure the star can produce. The iron core is compacted to a state that even overcomes electron degeneracy pressure. In this state of core collapse, electrons are forced into combination with protons so forming neutrons and the whole star undergoes a kind of instantaneous dynamic gravitational collapse. The collapsing envelope rebounds from the neutron degenerate core in the form of a supernova explosion which scatters the entire envelope including its metallicity far into the interstellar medium which it will both materially and gravitationally seed for birth of new stars. What is left is a degenerate neutron core. This is the densest physical and mathematically explainable state of matter in the universe. They generate unimaginably intense magnetic fields and can rotate at high speeds forming what are called Pulsars.

You will remember that the Sun is about a third of a million times the mass of the Earth. You will also remember that the white dwarf remnant of a Sun-like star has sixty percent of this immense mass compacted into a sphere only the size of the Earth. The degenerate neutron core of a star more than ten times the mass of the Sun is gravitationally compacted into a size equivalent to that of a small, Scottish, Hebridean island. A teaspoonful of neutron degenerate matter would weigh more than all of the battleships and land armaments of the past two world wars put together.

Stars greater than ten and less than fifty solar masses, (And there are some more than a hundred), also build up in layers. These stars are also subject to internal disturbance and near core element mixing caused by powerful convection currents. The gravitational energy of these stars is so unimaginably immense, their internal temperatures so high and their burning rates so fast that they only live for some tens of millions of years as opposed to the Sun's ten to fifteen billion years. These stars reach temperatures sufficiently high to fuse iron and the fusion of iron begins an immediate and cataclysmic sequence of events. Iron fusion, being an endothermic reaction, uses energy from the star. Energy is absorbed from the core which causes it and the immediately surrounding envelope to cool. This reduces the star's outward radiation pressure which, in turn, causes greater gravitational compaction. This increases the rate of iron fusion which uses more of the star's energy so that the whole becomes a self-feeding runaway reaction. Within a matter of seconds, an iron core, which may be tens of Sun masses, undergoes dynamic gravitational collapse. The envelope tends to follow and rebounds from it. The entire star explodes in a type ii supernova explosion. This is one of the most dynamically powerful events in the universe next to the Big Bang and, in its short duration, generates the immense temperatures and pressures that cause a kind of instantaneous ladder effect synthesising the whole range of elements heavier than iron.

What is left behind from a type ii supernova is very strange. All of the material of which the star's envelope was comprised, including its now comprehensive metallicity, has been blasted away into galactic sized regions of space. But there remains behind some strange kind of encapsulation of its core's gravitational energy. The core has collapsed even beyond neutron degeneracy into a state that is not yet understood by science. It is called a Black Hole. Black holes have been theorised by science over the past two centuries but have remained unobservable because their gravitational influence is so immense that even light cannot escape them. However, thanks to the Hubble space telescope, astronomers can now determine that they do exist and where they are. Most of them support the phenomenon mentioned earlier, Quasars, that

Hubble can see. It would seem that galaxies generally, including our Milky Way galaxy, have immense black holes at their centres; remnants of ancient stars that may have been millions or even tens of billions of solar masses. These stars, probably the class III mentioned earlier, would have had lifetimes of tens of millions of years rather than the ten to fifteen billion of our Sun and other younger stars. Some scientists believe the Black Hole remnant of these ancient, super massive stars may contain a kind of singularity similar to that from which our universe emerged in the Big Bang. As yet, we have no means of knowing the particular growth and death throes mechanism of such short lived and super massive stars. But Hubble can see that they have played a crucial role in the universal machine. They extend wide spheres of gravitational coherence in space into which they have spread the prepared material, set the pattern and act as focal points for the growth of galaxies of stars around them.

Chapter 10 - Philosophical Interpretations

Considering the foregoing sequence of events from the instant of the Big Bang, science, over the past few decades, has illuminated our universe, not as a hap-hazard, but rather, a carefully guided, intricately correlated, finely tuned, interdependent and specifically functional machine. To sum up, the instant of the Big Bang initiated a change of state from eternity to space-time. This was by introducing that strange entity we call energy. Energy motivates eventuality and space time correlates eventuality in the direction we call causation. Energy also incorporates the capacity to exist in different states; states that were specifically determined by a controlled process in the Big Bang event. This determined the fundamental forces which in turn determine the high degree of specificity apparent in the nature and structure of fundamental matter and its causational progression through space time in the structure and interaction of gross matter. Energy, in the form of inflation expanded the whole to form a locally inhomogeneous and large-scale homogeneous universe. Local inhomogeneity allowed formation of stars of varying masses. The variation in star masses allowed for stars sufficiently massive to produce metallicity but too short lived to allow for evolution of life systems. It also allowed for star masses insufficient to produce metallicity but capable of incorporating it and with lifetimes sufficiently long to allow for the evolution of complex life. The metallicity itself emerged from the star machinery richly endowed with its own highly specific chemo-mechanistic (Or quantum) relationships which we recognise as the science of chemistry and from which life emerged. In other words, the whole operates as a machine whose culminate output, in our universe, is intelligent life.

If an architect is designing a building or an engineer designing a structure or machine, it has to include a number of parameters. These are measurements, shapes, ratios, strengths, relationships etc. Our universe is ridden with close knit parameters. Science calls them The Universal Constants or Constants of Nature. There are a considerable number of these. They range from the speed of light to the basic unit of electrical charge to the configuration of the hydrogen atom to the force of gravity, electromagnetism, and the fundamental nuclear forces and so on. If any one of these values were different from what it is our universe wouldn't work in terms of producing life; it would be a spanner in the works. It would seem as though the prime energy at the first instant of the Big Bang was imbued with a code or a template that determined the structure and functionality of a universe with the potential to produce intelligent life. This forms the fundamental basis of the Anthropic Principle.

To modernise a classical, self-evident truth, if you were able to land on an extra-terrestrial planet and, during your exploration, you were to find a pocket watch or any other functional machine, you would not conjecture, you would KNOW that an intelligent being had left it there. You would know it was an artefact of intelligent design. You would know this because you know that inanimate matter does not gather, refine, shape and assemble itself into something that is intelligently and specifically functional. An artefact is a direct reflection of intelligence. The Anthropic Principle sees the finely tuned and intricately interrelated mechanistics of our universe in exactly the same light; it sees it as a direct reflection of a higher intelligence. It is an overwhelmingly powerful argument for the existence of God and has no counter argument. This is not religion, myth, magic or miracle; it is scientific observation. In this light, unless you can believe that a jumbo jet or a Large Hadron Collider could self- assemble by accident and without purpose you now have to believe that God exists. Or, as the astronomer Sir Fred Hoyle so aptly put it, "Either God exists or a hurricane blowing through a scrap yard could assemble a functional aircraft". Which is the more likely?

Previous to this revelation, one could concede some rationalisation in the choice of atheism over the myth and magic of biblical stories and, for some, the escape it offers from personal disciplines in terms of morality, human dignity, responsibility and purpose in our being.

Science is now in a phase of rapid learning of the intricacies of our universe and the more it learns the more it supports the Anthropic Principle. It is atheism that now dwells in the realm of magic and miracle in its belief that the incredible mechanistics and mathematical logic of our universe just popped out of nothing and in complete absence of any intellectual perception. Atheism is now a personal prejudiced choice; one without supportive argument and nothing more than a further conceding to convenience living. One thing it is not, is a higher intellectual view. Contrary to what the modern social writer might offer, it is just a softer, faith- based religion; one of denial and convenience rather than discipline. This is not to say that atheists are necessarily bad people. But they are blind to any sense of purpose in human being so, this makes it easier for them to be bad.

So, does this mean that God did indeed create the universe and man and give some insight to the major religions regarding his intent and purpose? No, it doesn't; such an idea makes no sense whatsoever. In his accommodation of science, the modern theologian has grabbed onto the Anthropic Principle and linked it with his belief in biblical style creationism. As a result, the principle has become culturally linked with religion so causing science generally to distance itself from it; even though it may be science's greatest revelation. In this respect, science is being subtly affected by religion in as much as, in its abandonment of the magic and miracle of religion, it has also abandoned the God idea. This, in turn, abandons any idea of purpose in human being. In our being, without purpose, there is nothing to which we might aspire and no basis upon which to build a globally coherent human society. In endowing us with intelligence, nature has given us the power to discern that there is purpose in our being and to rationalise what that purpose might be.

Science, over the past two hundred years and especially the past five decades, has considerably broadened our minds. It has enabled us to take a much more sophisticated view of the God idea. The greater our understanding of the universe in which we live, the greater becomes our understanding of the human disposition within it.

Whilst it may be absurdly naïve for us to question an omnipotent mind, we are, at the same time, intelligent beings. The question is therefore self- imposing. Why would an omnipotent, all powerful being want to create humanity? And if it did, why should it need to create and use a machine like our universe for the purpose? Why should it subject life to the raw, ruthless brutality of evolution by natural selection with such macabre life for life nutrition? (There are many animal species, from elephants to rabbits, clearly illustrating that murder of sentient life is not an essential feature of evolution).

However lesser our intelligence might be to that of an omnipotent being, our experience of intelligence tells us that our greatest mental need is plurality; the need to communicate and share intelligence. One of the worst human psychological tortures is prolonged isolation. Next to food and accommodation, media and communications are the widest sphere of human industry. Indeed, it was communication that induced one of the earliest uses of electricity as an applied science. Even science would be of little interest to many of those who call themselves scientists without its journals, seminars and other communicative access to the wider academia. Also, it is significant that, of the vast range of life species on our planet, the power of speech is attributable only to the one single intelligent species. In fact, the capacity to communicate intellectual reasoning is an innate and necessary feature of intelligence. On this basis, we can assume that if an omnipotent intelligence created us, this is the reason. However, this reasoning is not without serious problems. One can easily understand that an omnipotent intelligence must be alone. More than one omnipotent being is a contradiction in terms. But an omnipotent being would also be all that exists. If it created humanity then humanity would be of itself and therefore, no plurality. Human intelligence would just be a further, minor extension of the omnipotent intelligence. How then to arrive at plurality of intelligence without creating it? This is the Omnipotent Challenge.

Clearly, in terms of human logic, there is no absolute answer to this question other than to say, like a square circle, it is impossible. However, we can construe a scenario where the degree of divorce would be sufficiently and almost absolute. We shall consider this but it would be for an omnipotent intelligence to take it further and we shall see that in the nature of humanity, it has. "The Pearly Gates" are a barrier more formidable and the promise beyond them something more majestic than anything the bible ever dreamed.

At this point, I would beg the reader's indulgence as I intimate a short, if boring and seeming unrelated little story.

I have a nephew who runs a painting and decorating business. In what little spare time he has, he follows the rather peculiar pursuit of restoring old cars. This is purely a hobby and he meets with similarly minded people at various times and places where they congregate and show off the fruits of their labours. Four years ago, he was en-route to such a meeting in a nineteen forty-eight Jowett Javelin which he had spent some years restoring. Unfortunately, he was involved in a road accident where a large truck crossed the central reservation and ripped off half the front and near side front wheel of his Jowett. My nephew was hospitalised with broken ribs, a broken arm and a dislocated shoulder. To most people, this would have been a disaster. But to

204

habitual car restorers, it's an exciting new challenge. As soon as he was well, my nephew set about re-restoring his beloved Jowett. He managed this over a period of a year, all except the front wheel which had been irreparably deformed. He mounted a nationwide search for a nineteen forty-eight Jowett Javelin wheel but was unsuccessful. Some time later, he was attending a restored car meeting somewhere in central England. During the spread of fantastic mechanical stories which are a major feature of these events, he was informed of a scrap yard outside Birmingham which must have been about the only one he had missed. The following weekend, he visited this scrap yard and, after several hours of poking and searching, did, in some obscure corner, find what he was looking for. With a degree of delight only explainable to other old car restorers, my nephew was able to finish the re-restoration of his Jowett Javelin.

So, what is the point of this story? The point is the power of compatibility. The wheel in question may have lain lost for fifty years, being buried from time to time under other mountains of scrap and probably surviving removal because no one ever found the time to remove its worn-out tyre. But it had something going for it. It was compatible with a nineteen forty-eight Jowett Javelin car. It was the right width and the right diameter. It was formed to accept the right tyre. The fixing stud holes were set on the right periphery and at the right pitch and were the right diameter to suit the fixing studs on a nineteen forty-eight Jowet Javeline brake hub. These are the terms in which it was compatible just as the energy states of electrons in different atoms are compatible in terms of the molecules they form. That compatibility, however weak, however distant, formed a persuasion; a persuasion that was realised within an enormous sphere of pure chance. But that wheel was made for that particular model of car. Which car it would actually end up on was therefore a subject of pure chance emanating from a sphere of instituted chance. It is within this frame of conceptuality that the omnipotent challenge becomes, at least partially answerable, but how?

The reader, of whatever intellectual level, must now realise that I am in very serious difficulty. However, I do remain cognisant of the fact that I am attempting to answer what, in terms of human logic, is an unanswerable question. How can a being which is everything that can possibly be allow something else to come into being?

In the following, I shall propose a kind of scenario whereby our universe might have come into being. The reader should understand that this is a purely fictional imagining. None of it should be taken as real. Its purpose is merely to convey a conceptual frame within which one might perceive the possibility of an omnipotent intelligence achieving something which, to our limited intelligence, seems impossible.

We have already considered that our universe is a course of activity rather than substance and that what we see as substance is an effect of this activity. Just as the science of geology can retrospectively trace the evolution of our planet and palaeontology can retrospectively trace the evolution of life and astrophysics the evolution of our universe, so it should be possible to at least speculate on the most fundamental aspects of the activity of which the physicality of our universe is an expression.

We live in what one might call a bipolar universe. Everything is either on or off, in or out, up or down, light or dark, left or right, hot or cold, near or far, alive or inanimate, positive or negative, matter or antimatter and so on; all that is, except gravity. If we call what we experience as gravity positive gravity, we have no immediate experience of negative gravity. Also, of the four fundamental forces i.e. gravity, electromagnetism, weak nuclear force and strong nuclear force, science sees gravity as the weakest of these. But this is a purely relative notion. Gravity is a universal force whereas the other three are point related. We have already seen that gravity is the fundamental driving force of our entire universal machinery. In this respect, gravity could be considered the strongest force. Also, since the fundamentality of our universe is space-time related activity, i.e. everything that exists, exists by reason of its interaction with something else, then it is a fair assumption that negative gravity might exist. In other words, gravity, like everything else, is polarised. But normal, or what we are now going to call positive gravity, is an attractive, mass proximity related force as described by Newton's laws. We therefore have immediate experience of it. But negative gravity would be a repulsive, mass distance related force. Its value in our immediate presence would therefore be so close to zero we would not experience it. However, we would see its effect as Edwin Hubble saw it in distant receding galaxies. These galaxies are moving away from us and recently, Saul Perlmutter; a professor of physics at Berkeley California, demonstrated that they are moving away at accelerating speeds just as local bodies accelerate towards the Earth. Science attributes this peculiarity to some inexplicable form of dark energy.

We have already considered that the prime energy emanating from the singularity in the first instants of the Big Bang appears to have been imbued with a code of instructions for the assembly and functionality of our universe. The progress of the Big Bang from instant to instant and millennium to millennium appears as a precisely controlled and mechanistically interrelated series of events. But upon what and in what way is this code impressed? We have seen how the structure and functionality of life follows a code comprising molecular arrangements impressed on the structure of the DNA molecule. Our universe appears to be objective, even if only in the teleonomic sense, and its object appears to be intelligent life. It is a reasonable assumption therefore, that the code of life is a reflection, at least of the mode of coding, for the structure and functionality of our universe. Also, it is hardly conceivable that any number of different truly fundamental forces might accidentally combine to produce the highly co-ordinated and inter-dependant course of eventuality represented by our universe. It is much more likely that this arises from interaction between different facets of a single force. In this respect, and for the purposes of our little fictional scenario here, we are going to assume that the structure and functionality of our universe is coded for within the nature of interactive polarised gravity. We are also going to assume the nature of polarised gravity and say that it is the prime and strongest of the natural forces and that the polarised factions are co-transient. Co-transient means that each of the polarised factions has the capacity to change partly or completely into the other. (Co-transience might be a newly extended use of a word but it is not a new concept. We have seen, during the Big Bang event, how energy and matter are co-transient. Also, their degree of co-transience varies or is modulated by temperature, pressure and effects of the emerging fundamental forces changing over time during the course of the event and so producing the range of the sub-elemental constituents of gross matter. We have also seen how the immense power of gravity in the hearts of stars can re-convert this back to the energy state).

If gravity were to exist in this form and if the cycle of positive gravity were to be advanced in time by ninety degrees of its phase and that of negative gravity retarded by the same amount then the positive and negative factions would be directly phase opposed. (Phase shifting, both forward and backward is a common and indeed naturally occurring phenomenon caused by capacitance and inductance effects in electrical and electronic circuits). In the case of gravity, the phase shifting would be apparent and would be an effect of co-transience running at some ultimate frequency. (This could be what determines the rate at which time passes). If they were steady states and full phase opposed, they would cancel each other out completely. However, if the zero axis of co-transience were modulated, then any number of different partial cancellations could be affected. What we see as normal gravity may be an almost complete cancellation of the full power of prime gravity. By this means, the axis of gravitational co-transience, what I call the "Cotran", the most fundamental interacting activity in our universe, could carry a code for the energy-matter states determining the entire functionality of our universe; one single accident affecting a specificity on the co-transient states of energy and matter. But any information that is intelligently perceivable as a code is itself a reflection of intelligence. So how could all of this have come about if it were not created by God? It could come about if God created a situation wherein something, on the basis of pure chance, might or might not create itself and I believe this is the human disposition in the overall universal scheme. Contrary to Einstein's statement, "God does not play dice", I believe that our existence can only be explained on the basis that He does. Our universe emanated from an infinite realm of pure chance which emanated from an infinite realm of instituted chance; one in an infinity of falls of the pennies. I am cognisant of the fact that this is not absolute divorce from creation but it is as close as human logic can contrive. The teleonomic quality of our universe reflects the intelligence which instituted that realm of chance. We have already seen how the code for life is a self-assembling entity deriving of chance compatibilities and aligned to optimum effect by the self-induced principle of natural selection. Just as the thrower of our hundred pennies has neither control nor effect on the outcome of the throw, so it is with God and our universe. God did not create us or our universe. He allowed us to happen and, in terms of our ultimate purpose, humanity must create itself. In the light of this understanding we relinquish the participating idea of God and replace it with a more rational external motivator of self-imposed moral discipline and social responsibility.

In this respect, science can accept the God idea whilst still adhering to its purely naturalistic stance. This approach also answers many other seeming impossible questions like, for example, why life is subject to the cruel brutality of the evolutionary process and why individuals within the various spheres of life are subject to such apparent injustices. It explains why Mrs. Jones, a lone widow, lying in a hospital bed survives while Mrs. Smith with a large family circle and church community all praying for her doesn't. Neither we nor our universe are subject to any metaphysical interference whatsoever.

If God created a realm wherein every possible form of energy and or force field existed in every possible dimension but in a state of ultimate chaos then such a realm would contain an infinite sphere of compatibilities. It would be a matter of pure chance whether the persuasions represented by any particular set of compatibilities would be realised. If such a realisation did occur, then a set of logic would separate from the chaos. This situation could have existed for eternities of eternities and may have produced billions of different kinds of universes; some persistent and others reverting back to chaos. (The reader should note that I use the phrase "Eternities of eternities" merely as a figure of speech to denote times longer than the lifetimes of universes. Nothing happens in eternity). There may be many more lifeless universes outside our universe than there are lifeless stars within it. It would be a matter of pure chance whether one of these universes produced life. It would be a matter of wider chance whether such life produced intelligence and a matter of immensely wider chance still whether such intelligence might identify its disposition and implement the necessary disciplines in pursuance of its purpose. This last sphere of chance is the widest of all and may be that which makes the divorce of creator from created absolute. An intelligence at or about the stage of present humanity would know that, other than the grace of living the social dream, there is no immediate heavenly reward for its effort. There is no such thing as individual, personal salvation. There is no ethereal, eternal, heavenly or hellish state into which we pass in an afterlife. Humanity at this stage, like the first microbes, is but evolutionary fodder in the omnipotent scheme. Yet, in this understanding, humanity must assume a dignity and, with discipline, pursue its purpose by way of social, moral and intellectual development. Humanity must survive itself. This would be a foot on the first rung of the ladder to Godliness. Somewhere on that ladder disease and death would be conquered either by natural course of evolution or by technological development or an intertwining of both. We have seen how the gene transmits instinctive behaviourism the overall

210

ecological effect of which is hardly distinguishable from intelligence. It would be reasonable to assume that one day it may transmit conscious memory so imparting its own immortality to its ultimate product. Intelligence and death would seem to be quite incompatible. Bunkum you might think but try to imagine the biotechnical leap from chemically induced instinctive behaviourism to self-conscientiousness and imaginative thinking. This was a leap from a bee to an Einstein. It is only that it happened that makes the immensity of it even remotely imaginable. As we have already seen, imaginatively thinking humanity has existed for about forty thousand years and we now consider ourselves to be intellectually and technologically advanced. But it has taken more than three and a half billion years to reach this stage and our planet yet has a potential life of another four to five billion years. Though we have advanced little or nothing on the social scale, (Science is the only coherent group of thinkers on our planet), we stand in awe of our technological advancement over the past two hundred years. Where shall we be after another four thousand years or forty thousand years or after another forty million? These are less than eye winks in evolutionary and cosmological time. It is as incomprehensible to us as were we to the first microbes. As a species, and providing we can survive ourselves, we could reach a point on the ladder where we commune both sensually and intellectually with the omnipotent intelligence and share infinitely eventful experience in an infinity of time.

I would here reiterate and advise the reader in the strongest possible terms that the foregoing Cotran scenario is a purely fictitious construction. It may be consistent with Newtonian gravity but it is hardly consistent with Einstein's theory of gravity. At the same time, I have no doubt that our universe came into being by means of a similar kind of scenario; the chance derivation of a fundamental set of energy states where compatibilities formed persuasions whose realisation is a functional universe reflective of but separate from an omnipotent institutionalising intelligence. Going back to our one hundred pennies, no matter how many times

211

they are tossed, even over eternities of eternities, no matter the number of patterns in which they fall, none of them are either ordered or disordered in absence of a perceiving intelligence

.

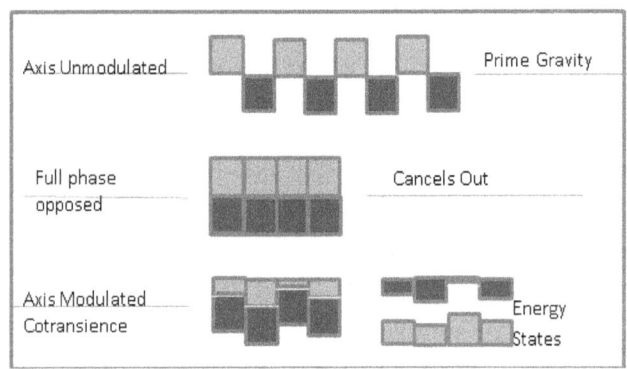

[OBJ]Fig. 27 Axis modulated co-transient gravity

Just as a matter of interest in passing, you may note from Fig. 27 that co-transient gravity could produce two codes so perhaps we have two universes operating inside one another. Also, at about half the current age of our universe, some six and a half billion years ago, the rate of expansion expressed upon it by inflationary influence had begun to slow down. It seems as though gravity was making a grab for it in an attempt to pull it back. But then the expansion began to speed up again. This would seem as though positive gravity hadn't made its grab quite soon enough and the universe had reached spatial dimensions where repulsive, mass distance related negative gravity could take over. Also, it is conceivable that if the Cotran scenario were true, there could be a couple of blips on the code. These could cause our universe once in say twenty or thirty billion years to be subjected to the full power of prime, positive gravity. This would crunch our universe back to a singularity. It may then be subjected to a pulse of the full power of prime negative gravity causing it to re-inflate into a new working universe; an eternally cycling machine awaiting fulfilment of its purpose. The Big Bang, inflationary theory and universal expansion describe, at least, three aspects of that scenario.

Regarding the inconsistency of our fictitious Cotran scenario with Einstein's theory of gravity, this theory also has its problems. I would now have problems with GCSE level maths so it's not for me to question Einstein. However, I feel there might be some problem on the part of the scientific academia in conveying Einstein's ideas to the public at large. For example, rather than gravity being an innate feature of mass, according to interpretations of Einstein's theory, space bends in the presence of mass and this bending of space is gravity. The classical analogy for this is to consider a horizontally suspended sheet of flexible material with a heavy ball bearing placed in the centre. The mass, or weight, of the ball bearing will cause the flexible sheet to stretch downwards forming a conical dwell in the sheet. Apparently, mass affects space in a similar manner and this bending of space is gravity. Fig 28 is a classical illustration of this.

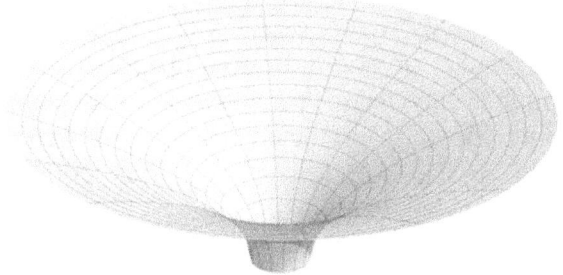

Courtesy Allen McC CC By – SA3.0 Via Wikimedia

Fig 28 Bending of space by mass. What I call the trumpeting effect.

But how do we reconcile this with an isotropic universe? In an isotropic universe, surely the effect of mass on space would have to be multilateral. If you try to consider the effect shown in Fig. 28 on a multilateral basis, you start by getting something like Fig. 29 and soon, the shape becomes both incomprehensible and impossible to illustrate. Would it be more correct to say space implodes in the

213

presence of mass? This would cause the same gravitational lensing as that observed by Sir Arthur Eddington during the 1919 total solar eclipse. This also would be difficult, if indeed possible, to illustrate, as attempted in Fig. 30 but it fits better with our intuitive thinking; not because it's any more comprehensible but because we experience it every time we lift a cup of coffee or a bag of garden compost, be it in London or Sydney. It would also be consistent with an isotropic universe. Also, there is the question: what other inherent property of mass is there that interacts with space in such a way as to bend it and cause gravity? Is it possible there could be a little "Cart before the horse" here? It is also interesting to note that whilst galaxies should present the ideal opportunity to witness this trumpeting of space, no such effect is visible in galaxy shapes. Galaxies have unimaginable mass and they stretch for hundreds of light years across space. They have high mass density at their centres and possibly even a black hole; yet, they show no sign of this trumpeting effect on space. They exist essentially as rotating discs, they have a bulbous, high mass centre and spin on a single plane. There are other shapes of galaxies ranging from oval, or egg-shaped to irregular or non-definitive shapes but none displaying space shaping in any form. Yet, there is no question, space does bend. Albert Einstein predicted it and Sir Arthur Eddington demonstrated it. The question is, is it trumpeting or multilateral implosion? And, what causes it?

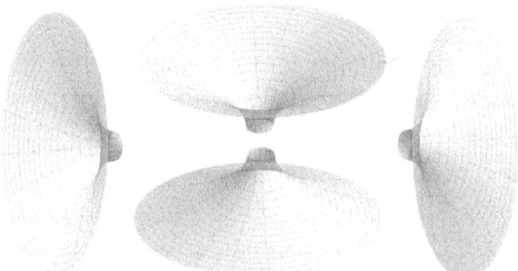

Images courtesy Allen McC CC by – SA3.0 via Wikimedia
Fig. 29 Attempt at multilateral bent space.

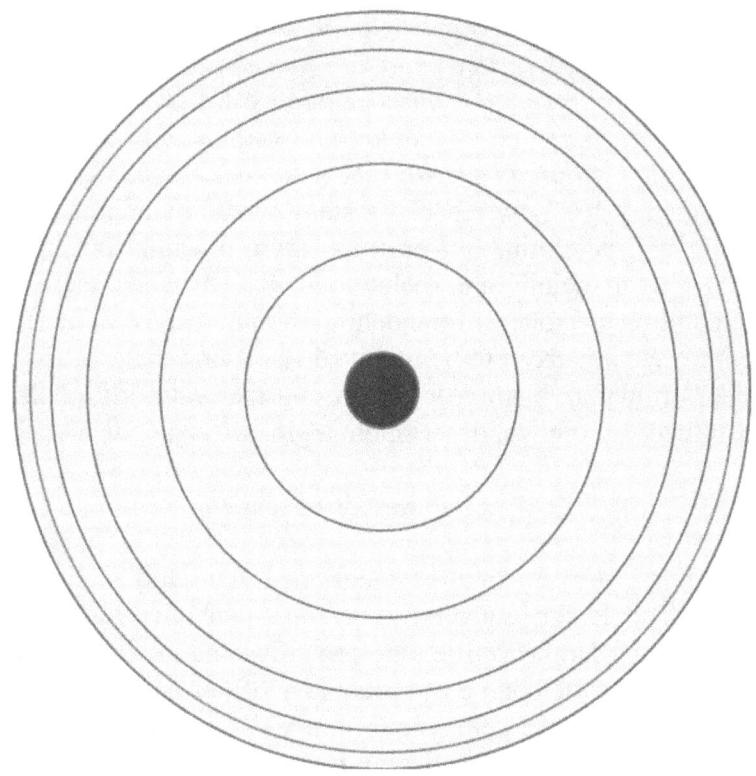

Fig. 30. Spherical, multilateral, imploded space: note the closer to the mass, the greater the amount by which space is stretched.

I write the following paragraph in italic since I may refer to it from time to time as "The Basic Question".

Of course, you may consider the foregoing to be a load of pure bunkum. But the question still remains. Which offers the better prognosis for human social and intellectual development? Is it the idea that our being is hapless, without meaning and leading nowhere; that our social development depends upon imposition of law rather than empathy and moral conscience; that there is nothing against which we can sin; that our sense of human dignity is merely one of fashion? Or is it the idea that there is a higher purpose in our being; a majesty to which we might aspire in terms of empathy, social justice, honour, moral dignity, freedom and intellectual development; a purpose against which we can sin or to which we can contribute; the idea that this humanity might identify its disposition within an overall scheme of meaningful existence?

Humanity has now reached a stage where its approach to this question will determine not only its future potential but its future existence as a socially and technologically advanced society. The grandchildren of today's grandchildren may enjoy or suffer the outcome. We are about to be involved in solving the greatest social problem in human history or of entering the reality of a hell more hellish than any biblical description, a global Nazi style holocaust.

Pyramids of fear.

Earlier, we very briefly mentioned the Drake equation in passing. The Drake equation was constructed by the American scientist Frank Drake during nineteen fifty-nine - sixty. Its purpose was to estimate the degree of probability of our being contacted by an extra-terrestrial intelligence. Frank Drake and the famous astrophysicist, the late Carl Sagan, founded the S.E.T.I. (Search for Extra-terrestrial Intelligence) institute in nineteen eighty-four. S.E.T.I., with an increasing range of highly sophisticated equipment has been searching for E.T. ever since without a single morsel of result. At the time the equation was constructed, none of its terms were quantifiable. Nevertheless, Frank Drake made some estimates resulting in the equation suggesting some two thousand intelligent species in our Milky Way galaxy.

The terms of the equation are as follows: -

$N = R^* . Fp . Ne . Fl . Fi . Fc . L$

Where: -

N = the number of civilisations in our galaxy with which radio communication might be possible;

216

And

R^* = the average rate of star formation in our galaxy

Fp = the fraction of those stars that have planets

Ne = the average number of planets that could potentially support life per star that has planets

Fl = the fraction of those planets that could support life that actually develop life

Fi = the fraction of planets with life that actually go on to develop intelligent life

Fc = the fraction of civilisations that develop a technology capable of sending radio signals

L = the length of time over which such civilisations release detectable signals into space

Over the past fifty years and especially the last two decades, the terms Fp and Ne on the right side of the equation have become, if not accurately, at least satisfyingly calculable. We now know that stars with planets are more the rule than the exception. (Indeed, in the light of what we have already considered, this might have been predicted from a proper understanding of the Anthropic Principle). It may be a very long time before we can quantify any of the remaining terms of the Drake Equation, if ever, except for the final term L. Throughout the whole entirety of our universe we have one single subject of study for elucidation of the value of L. That is ourselves; this humanity and, on this basis, we may know the value of L within the next one hundred years. For the sake of our grandchildren's grandchildren let's hope it will remain as ambiguous as the others.

It is some two hundred years since early modern science began to recognise the existence and importance of our planet's highly integrated ecological systems. Since then, ecology has become an important science on its own and scientists have ingeniously found or innovatively constructed methods of studying Earth's past ecologies. They can tell us about probable causes and factual effects of volcanism, plate tectonics, meteor impacts, ice ages, global warming and so on. But until two hundred years ago, there was never any sign of humanity having had any noticeable impact on planetary ecology. During its forty thousand years up until that time, human populations generally, like all other life on the planet, lived on or just below subsistence level. Except that farming as opposed to hunter gathering had given humanity a considerable edge, his life expectancy and procreation levels were, for the most part, still controlled by the same forces affecting all other animal life. In other words, humanity was still ecologically compatible. However, firstly, colonisation of the African, Indian and American continents then the industrial revolution, starting in the United Kingdom, and spreading explosively throughout Europe, was the beginning of a drastic sea change in this delicate and complex ecological balance.

Though it had been practised for some centuries by warring emperors, kings and nobles, the shipping industry then the industrial revolution saw the real, widely acceptable establishment of capitalist investment economics. This facilitated a rapid growth in innovative engineering and related sciences as well as extensive distributive and communicative infrastructures. Farming became largely mechanised and there was large scale migration of human resources from the land to industrialising cities. Industrialisation stimulated a slight lift onto or slightly above subsistence level living for large volumes of people and city density of population coupled with greater demand for human resources increased rates of procreation. An unprecedented rise in population began and this has been exacerbated by great advances in medicine altering the naturally imposed birth to death rate ratio. This together with an ever-increasing demand to move from subsistence to convenience living further fuelled and widened spheres of industrialisation and this has persisted to this day. Through the trading of goods and artefacts for delicacies and resources beginning mainly with textiles, spices and drugs, countries remote from the industrial centres, the so-called underdeveloped countries have been, to some extent, similarly affected. Indeed, and ironically, the highest population growth rates percentage of populations are presently in the now rapidly developing countries.

Courtesy Wikimedia commons
Mountains of waste to landfill

Over the past four decades or so, there have been many scientific and intergovernmental seminars and debates arriving at various agreements and disagreements regarding methods of dealing with planet-wide pollution and its effects on climate change. They have amounted to little more than tinkering. This is because it would be politically and career disastrous for any politician or scientist to even mention the real crux of the problem; that being, **there are far too many people consuming far too much and producing far too much waste.**

[OBJ]

219

Waste vomiting from our oceans and everything that lives in them. A visible planetary melanoma illustrating the unsustainable cost of mass human convenience.

Masses of planetary resources and potential, non-degradable waste encapsulating each individual driver.

Courtesy Wikimedia commons
Masses of people

Crowded dwellings

Convenience living is an evolving process and, as such, cannot be turned back. But it has a cost. It cannot run parallel with ever increasing population. One example of this is the present human demand for convenient, cheap, tasty protein. There are now some one and a half billion cattle being farmed on our planet to meet this demand. The pollutant effects of this are enormous. Cattle and their management produce nearly twenty per cent of the greenhouse gasses that cause global warming. That's more than all of the world's cars and other land, air and sea transport systems put together. The feed, waste, fertiliser and general farm/ranch management systems associated with cattle are lethally pollutant. The flatulence of the animals themselves produces thirty-five per cent of the world's methane emissions and this is two dozen times more lethal than carbon dioxide. Their feed fertiliser, waste and waste management produce sixty-five per cent of human related nitrous oxide which is three hundred times more pollutant than carbon dioxide. There are tens of thousands of square kilometres of off shore sea beds which, only thirty years ago, were havens of coral supporting myriad life forms. Now these are completely dead and barren due to cattle management waste issuing from rivers into the sea. Tens of thousands of square kilometres of vital forest lands are being destroyed to provide cattle grazing land and just as much is being turned into arid desert land from over grazing. All in all, cattle are responsible for more than seven billion tonnes equivalent CO_2 emissions per year into our atmosphere. This is only one of a host of human related pollutants affecting every aspect of our environment and the effects of which are estimated to double or even treble over the next thirty years.

Our world is therefore no longer populated by humanity; it is lethally infested. We have allowed humanity to become a widespread, deep rooted and aggressive cancer disrupting the integrity of our finely balanced global ecology. Like the mindless little microbes that turned our planet into a frozen ball of ice for three hundred million years, humanity is now rapidly breeding itself into imminent extinction

Cities towering upwards to save land space

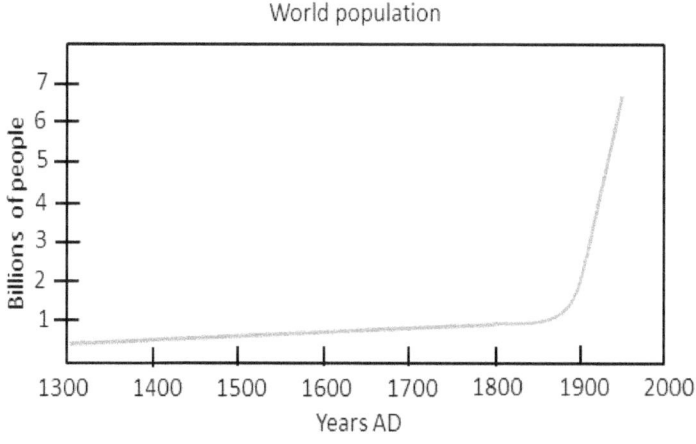

Fig.31 Population graph 1

You will, no doubt, find Fig. 31 a rather alarming revelation but it is how things actually are. You can see from it that, by only seven hundred years ago, and having existed for some two million years in all, global population by humanity had reached only some three hundred million. Over the ensuing six hundred years, till the time of the industrial revolution, it tripled to one billion. Then, over only the next one hundred years, it doubled to two billion and started its glutinous consumption of fossil fuels and other mineral resources. Over only the past sixty years, world population has more than doubled from three billion to getting on for seven and a half billion. It is heading for upwards of fifteen billion by the end of this century.

Together with that, three quarters of the world's existing population are only now beginning to express their demand on planetary resources and place their footprint in the already suffocating well of polluted planetary pus. It has taken the western world (About one quarter of the world's population) some one hundred and sixty years (Since about the time of the industrial revolution) to reach its present state of convenience living. The remaining three quarters of the world's population will do the same over the next twenty years. Even without its predicted population increase of another two billion, our demand on the planet in only twenty years' time will be four times greater than it is now. Over the past fifty years alone we have consumed more of our planet's resources per day than we did in a generation at the time of the industrial revolution.

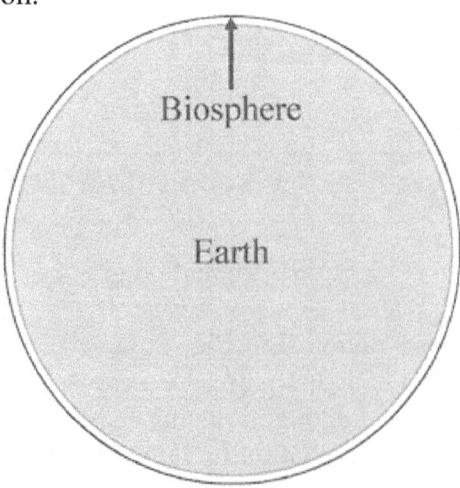

Fig. 32. The biosphere.

In fig. 32, the small gap surrounding our planet represents the biosphere. It includes the atmosphere, the mountains, the top-soil, the sub-soil and depths of the oceans. If you were to compare it with an apple skin, the skin would need to be twenty times thinner than that of an ordinary apple. It is within this complex, fragile, delicately balanced sliver of haze that all known life in our universe exists. It is now beginning to fail under relentless, ruthless, human assault and we are failing completely to make any serious attempt at its management.

Clearly, this situation cannot be allowed to continue. Our planet, like any other vessel, can only contain and sustain a certain maximum number of people and maintain integrity of its fragile and complex ecological balance. The question needs to be addressed urgently and the prospects of its implications are desperately frightening. By the end of this century, we need to reduce our planet's population to what it was a hundred years ago, at around one and a half billion and continue from there towards a target of one billion. Measures would have to be taken on a global scale that would threaten governments; democratic or otherwise. Measures like devising reducing rather than growing economies and controlling birth rates through a globally, coherent, compulsory mass neutering campaign including all reproductive genders. These may seem like mind-bendingly horrific measures, but we may already have entered an insidiously developing, irreversible, apocalyptical state. We are playing for the continued existence of this humanity. The population level on our planet, which is driven by the most compulsive power in nature, can no longer be left to individual personal whim. It has to be strictly and effectively controlled. Over the past four decades, a birth control policy in China, only very loosely applied, has prevented nearly half a billion births. It's a pity indeed that the world as a whole did not read the same message forty years ago.

Controlled population reduction would be an extremely complex scenario due to demographic differences in carbon footprints although this would tend towards some levelling as
225

populations were reduced. For example, a single person's carbon footprint in the UK is about three times higher than that of a person's in India and a person's in America is about three times that of a person's in the UK. Also, the very rich, who have the highest carbon footprint of all, would be amongst the first to feel the effects of reducing economies and would combine in a most powerful lobby against it. Together with this, current thinking within the social academia is leaning towards educating women in family planning as a means of population control. This is a clear indication that we are nowhere near the level of thinking needed to appreciate the imminence, the immensity or apocalyptic nature of what is descending upon us. Immediately imposed population control is the only tool at our disposal that can have any effect on the impending climate change crisis within the time hopefully available to affect it. There is a point in time where ruinous climate change will become self- sustaining and unstoppable. We don't know exactly when that will be, but have to hope we can reverse the trend before it happens. And, in order to do this, we shall require something, as yet, never achieved in human history; globally coherent governmental consensus.

Whilst requiring superhuman thinking and superhuman effort to affect, the foregoing is our only practical means of arresting climate change. However, there is, currently, a revival of Godless forces at work in our world who would hinder or frustrate this in favour of a more typically human approach.

Secularism, in the literal sense of the word, would seem about the soundest basis on which to build a modern democratic society. Unfortunately, however, it is bugged by a serious problem. It is a purely individualistic stance and therefore has no communal rule of practice. (Even science has communal global coherence determined by its mathematics and a cultural infrangibility in the truths it accepts. Scientists throughout the world can talk to one another.) As a consequence, secularism tends to deteriorate into atheism and loses all sight of ultimate human purpose.

Whilst the top Nazis in early nineteen-thirties Germany were of Christian backgrounds and presented themselves publicly as such, they forsook that ideal and became atheists (Except that Hitler himself began to nurse some ideas of personal Devine association and other top Nazis descended in groups into the pits of occult thinking). As the Nazi ideal grew in popularity throughout that decade so did the amount of status related regalia attached to it. This had an effect of generating strong narcissistic tendencies and this, in turn, gave them the empathetic insulation and consequent moral freedom necessary to accommodate the holocaust.

The somewhat unimposing person of Adolf Hitler utilised the economic downfall of the nineteen-thirties to gain a degree of political celebrity status. Whilst the economic crash was global, Hitler convinced the German people that their plight was a direct result of impositions, especially the "War Gilt Clause" imposed on them by the Treaty of Versailles because of Germany's initiating the First World War. He had a degree of oratory skill which he used to deny Germany's guilt and rouse the people into a new sense of nationalist pride. He was telling the people exactly what they wanted to hear and, on this basis, he cultivated a magnetic charisma. This engendered a degree of popularity sufficient even to reconcile a number of already opposing political factions and eventually elect him to a position of unassailable and despotic leadership. He became the insane and irremovable pinnacle of a pyramid of fear. This spread downwards and outwards through his henchmen and subordinates and through the military and civil institutions and into the populace of Germany itself. Any misalignment of ideas from those of the despot became a serious threat to one's life and indeed many thousands suffered imprisonment, torture and death for trifling or even foundless alleged differences. It was a situation where man and wife, father and son, brother and sister and even neighbours bore elements of fear and distrust.

Hitler had an innate hatred of the Jews born of nothing but the cultural effects of having grown up in Austria where Jews had

227

been welcomed and rejected several times throughout history. He was rather simple minded and consequently had no empathy towards people or anything else. He murdered his own faithful dog just to test a poison. He allowed his hatred of the Jews to become an insane obsession and there was one particular facet of life that he well understood i.e. crowd psychology. With this understanding and the administrative help of his henchman Joseph Goebbels, he built a powerful propaganda machine which he used to radicalise the entire German nation into the Nazi way of thinking (It is important to note and remember that we are all subject to media persuasion and media power in Hitler's time was miniscule in comparison to that of today). The subsequent mind crippling fear, confinement, forced movement, assunderment of families, physical abuse, torture, enslavement and mass murder suffered by the Jews was a hell beyond any words to describe. Atheist Europe, including the United Kingdom, is presently seething with hard left, hard right and fascist activists who would revel in repeating this whole scenario.

Courtesy Wikimedia commons
The previous week's business at Belsen concentration camp northern Germany as found by the advancing British eleventh armoured division, April 1945; pit 1 of 3 containing the unburied corpses of thirteen thousand murdered Jews and Russians.

The point of all this is that Hitler initiated the holocaust on the basis of nothing more than personal whim. In fact, it considerably weakened his Germany's chances of winning the war. But consider the situation now when a despot might offer a seemingly genuine reason for mass annihilation on the basis of gross overpopulation. If this were to happen and proceed, and there is an increasingly threatening possibility that it could, the repercussions would be such that this humanity would rapidly descend into a state of anarchy from which it would never recover as an advanced social, intellectual and technological civilisation. It would establish Drake's value of L for this planet at some two to three hundred years. Considering it took some four billion years to evolve to the point where we could search for ET and if our value of L is confirmed and typical, the probability of two separate planetary civilisations overlapping in time and within communicable distance is so remote as to be out-with any reasonable probability.

Throughout human history the only significant steps towards a globally coherent society have been the federalisation of the American states and the formation of the European Union. The European Union was formed in order to prevent a repeat of Hitler's Germany. Unfortunately, because we now happen to be in difficult economic times and the European Union is still suffering its growing pains, there are activists taking advantage of this and utilising the crowd mentality with the exact same rhetoric as Hitler did to re-introduce nationalist ideas and promote ideas of indigenous right together with cultural and religious division in an ever-shrinking world. The western world as a whole is now moving towards hard right politics and this is but a bird-hop from Fascism which, in turn, is but a bird hop from Nazism.

A major step towards global social collapse and re-introduction of jack-boot tyranny was successfully engineered in the UK in 2016. It was achieved mainly by two charismatic, hard right activists in the Euro referendum utilising the current, widespread fear of immigration. They got away with it by having an almost unopposed free run in the lead-up campaign and addressing the trades and services mass of our population with blatant disinformation. As politicians, the leave campaigners knew full well that free movement of people, goods and services is a fundamental tenet of the European Union and there will be no trade deals without it. Yet they hoodwinked the trades and services mass of our population into believing otherwise. Who would ever have believed that it would be the United Kingdom that would be duped into moving politically so far right as to undermine the only serious trend towards global social coherence in the modern world?

This shift towards hard right politics in the United Kingdom gives credence and tremendous stimulus to the seething hard right activists throughout Europe, the UK and America. It could be the instrument for the breakdown of the union altogether. It could cause a regression of the social state of our world by a hundred years; a hundred years we just cannot afford.

Humanity has as yet failed to establish any globally coherent society and is about to face the greatest social problem in its history; one that can only be dealt with on a global basis. Human infestation of our planet has to be addressed and urgently before our planet tips into irreversible ecological collapse in terms of its capacity to support intelligent life.

Courtesy Wikimedia commons
Humanity may already be in its final hours

230

The ways in which human infestation are affecting our planet are myriad, subtle and imminently lethal. It would take some book volumes to enumerate the causes and effects but we are ravaging our planet as though there were no tomorrow. Unless we can find ways of dealing with the problem now, there will indeed be few tomorrows as we know them.

The question now is how can this looming crisis be dealt with? Can it be dealt with? Or have we already passed a point of no return? There are several possible scenarios. One is, we continue to ignore or fail to recognise the enormity and imminence of the problem and continue to let nature take its own course. In that case, by the middle of the next century, most or all of the world's coastal cities will have been lost to the sea and this may already be unavoidable. Huge areas of agricultural land will be water sodden and rotting into peat bog. Even large areas of the higher topographies will be flooded from persistent torrential rainfall. Large scale migration of populations seeking more tolerable climates will have begun. This will result in endless warring, the acceptance of the viability of mass annihilation and the beginning of humanity's irreversible descent into a state of anarchy. Before the end of the present century, one of the world's greatest financial and resource commitments will be in defence of sea invasion and even this can only be temporary.

There are other scenarios but unfortunately, in our present state of global society, with its nationalist and ethnic divisions, conflicting political and religious dogmas, political and corporate corruption and unjust wealth distribution they only get worse and worse. They all lead in one way or another to ascendancy of despotic, dictatorial regimes, mass annihilation and descent into irreversible anarchy. This humanity is approaching the brink of failure in terms of its ultimate purpose; a purpose it has yet to identify.

Yes, it is a doom and gloom scenario but not necessarily entirely so. There might, just might be sufficient time to deal with the problem in a civilised manner. Even this however will be

231

extremely painful and seemingly grossly unjust to many. However, we must attune our minds to the simple fact that a controlled prevention of life is much more acceptable than the mind-bending fear and ruthless murder of the living.

Referring back to the "Basic Question" you can see that there are two lights in which the problem might be approached. In the former, the problem doesn't really exist as such. In this light, humanity and our universe are, ultimately, hapless, meaningless events; the destruction of humanity just wouldn't matter. We will have enjoyed or suffered what we had of it and that's it. Our disappearance would be of no more significance than ridding a nest of bed bugs. In the latter, humanity has a purpose, the potential majesty of which is, as yet, beyond our power to comprehend. And we may have a value which is immeasurable. We may not only be unique in our own universe. We may be unique in billions of universes or even universes of universes throwing the pennies over eternities of eternities; a degree of chance matched only by the enormity of the promise.

And the promise doesn't come from Devine intuition or soothsaying or magic or miracle or any other hocus-pocus. It comes from the dedicated, hardworking, skilful and ingeniously innovative empiricism of modern science. The Anthropic Principle is clearly written in the incredible mechanistics and logic of our universe. God, whilst being quite alien to the comfortable, forgiving negotiator described in biblical stories, does exist. Can this humanity rise to that realisation? In dealing with the population crisis, we shall become both the judges and the judged.

Heaven and Hell are not God-made states. Heaven is humanity's recognition and achievement of his purpose. Hell is his failure and the indescribable holocaust of his way to oblivion by his own hand.

In early 1940, just after the beginning of world war two, the civilised world sat back, sighed and asked itself "How on earth could we ever have let this happen?" This, even though the writing had been clearly on the wall for a decade before, just as it is now.

In all of human history, our world has never been in such a critical situation as it is at present. We are looking self-imposed extinction straight in the face; and this whilst the vital political time and energy of the western world is being gluttonously sapped in its grappling its way through a far-right political, economic and social sink-hole of social fragmentation and nationalist isolationism; the social insanity of Brexit.

The Large Hadron Collider must be about the most complex machine on Earth. As an artefact, it is a fine reflection of human intelligence. If the scientific community would but see its own revelations of the incredible mechanistics of our universe as a similar reflection of a higher intelligence in terms of the Anthropic Principle this would change our climate of thinking and this humanity would at last have recognised its purpose. Perhaps then, even at this critical hour, we might yet save our world and lay the way to achievement of that purpose.

There is no yardstick of measure, no words to describe and no breadth of imaginative mind to comprehend either the fulfilment or the frustration of human purpose; to define Heaven or Hell.

Population graph 2

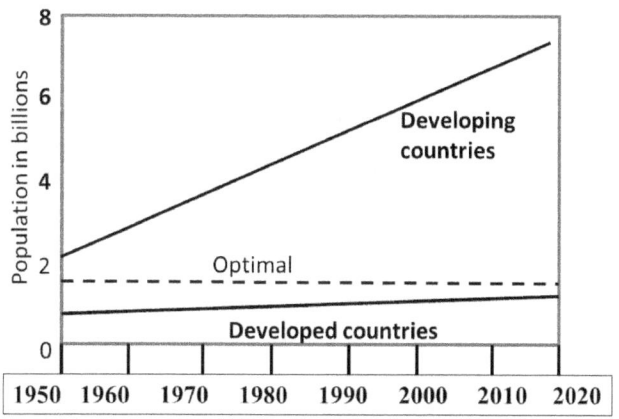

In Population graph 2, you can see the world population growth rates over only the past seventy years. The Lower slightly angled line shows the population level in the developed countries in 1950 of eight hundred thousand million and rising to one billion today. The upper steeper line shows the population level of two and a half billion by adding the developing countries in 1950 and rising to almost eight billion today. The dashed line shows the optimal global population level that is compatible with our universally unique ecological machinery. In only twenty years' time, the global population will be approaching ten billion and what we now call the developing countries will be fully developed in terms of their resource demands, waste production and ecological effects.

Our scientific academia is well used nowadays to thinking in terms of enormous time spans; cosmological, geological and evolutionary. That a disastrous planetary event could arise and insidiously overtake them within a single human lifetime has taken them somewhat by surprise, but here we are in the midst of it. There are hardly words to describe the enormity of our ecological problem or indeed the degree of human effort required to solve it. No amount of tinkering with solar panels or windmills and the like can have any measurable effect in the time hopefully available. Only a globally coherent and massive effort overriding all emotional, religious, political and dogmatic persuasions to get world population onto the black dashed line, in Population graph 2, by the end of this century can hope to solve the problem. We either achieve this or, we lose our planet and, in the most literal sense of the words, **go to hell** in the process.

Epilogue

Though we look and search and listen for the merest peep or bleep
of sound
Mindless silence still pervades our vast and starred surround
Lest we understand from what we came and this time round can
change the game
Our silence too must surely soon and for ever more abound

ACKNOWLEDGEMENTS

Wikipedia.

Wikimedia commons.

Jim Clark, Chemguide.

My loving and ever-caring wife, Edith for her forbearance.

236

INDEX

www.ingramcontent.com/pod-product-compliance
Lightning Source LLC
Chambersburg PA
CBHW071416180526

45170CB00001B/121